T0132630

Optimal Resource Allocation for Distributed Video Communication

Multimedia Computing, Communication and Intelligence
Series Editors: Chang Wen Chen and Shiguo Lian

Effective Surveillance for Homeland Security: Balancing Technology and Social Issues
Edited by Francesco Flammini, Roberto Setola, and Giorgio Franceschetti
ISBN: 978-1-4398-8324-2

Music Emotion Recognition
Yi-Hsuan Yang and Homer H. Chen
ISBN: 978-1-4398-5046-6

Optimal Resource Allocation for Distributed Video and Multimedia Communications
Yifeng He, Ling Guan, and Wenwu Zhu
ISBN: 978-1-4398-7514-8

TV Content Analysis: Techniques and Applications
Edited by Yiannis Kompatsiaris, Bernard Merialdo, and Shiguo Lian
ISBN: 978-1-4398-5560-7

Optimal Resource Allocation for Distributed Video Communication

Yifeng He ◆ Ling Guan ◆ Wenwu Zhu

CRC Press
Taylor & Francis Group
Boca Raton   London   New York

CRC Press is an imprint of the
Taylor & Francis Group, an **Informa** business
AN AUERBACH BOOK

CRC Press
Taylor & Francis Group
6000 Broken Sound Parkway NW, Suite 300
Boca Raton, FL 33487-2742

© 2014 by Taylor & Francis Group, LLC
CRC Press is an imprint of Taylor & Francis Group, an Informa business

No claim to original U.S. Government works

Printed on acid-free paper
Version Date: 20130507

International Standard Book Number-13: 978-1-4398-7514-8 (Hardback)

Visit the Taylor & Francis Web site at
http://www.taylorandfrancis.com

and the CRC Press Web site at
http://www.crcpress.com

Contents

List of Figures

List of Tables

Preface

Recent years have witnessed a dramatic growth in multimedia applications, due to the tremendous advancement in several emerging technologies in digital media, communications, and networking. Multimedia data in a variety of popular formats such as image, animation, audio, and video have become the mainstream contents on the Internet and are being accessed by various mobile devices through wireless links. The diverse deployment of multimedia applications has been driving the field of multimedia communications and networking into a new era of truly ubiquitous media consumption, at any time and from anywhere.

Many multimedia applications, such as large-scale IPTV and collaborative wireless video streaming, involve real-time video transmissions over distributed systems. Examples of such distributed systems are Peer-to-Peer (P2P) streaming systems, wireless ad hoc networks, and Wireless Sensor Networks (WSNs). Though each type of distributed system has its own features, they share two common characteristics as follows. 1) Each node only knows about its neighbors, and does not have global knowledge. 2) There is typically no centralized controller, who can coordinate the behaviors of all the nodes.

There are many challenges for real-time transmissions of video over distributed systems. First, video streaming applications are sensitive to packet delay and packet loss. Second, network conditions and characteristics, such as bandwidth, packet loss ratio, delay, and delay jitter, vary from time to time. Third, it is challenging for the underlying networks (e.g., Internet and wireless networks) to provide real-time Quality of Service (QoS) guarantees to video streaming applications. Optimal resource allocation provides an efficient solution to improve the video quality for video communications over distributed systems. The problem in the distributed systems can be formulated into a resource allocation problem with an objective to maximize (or minimize) a performance metric, subject to the resource constraints at each node. Since there is no centralized controller in the distributed systems, a distributed algorithm is the desired solution in terms of the scalability and the communication overhead.

This book examines the techniques for optimal resource allocation for distributed video and multimedia communications over Internet, wireless cellular networks, wireless ad hoc networks, and wireless sensor networks. This book consists of six chapters. **Chapter 1** provides an overview of distributed systems, the challenges for distributed video communications, and the recent advances on optimal resource allocation for distributed video communications. **Chapter 2** presents the resource allocation techniques for scalable video communications over the Internet or wireless networks, including the following three specific topics: 1) network-adaptive resource allocation for scalable video streaming over the Internet, 2) Quality of Service (QoS) adaptive resource allocation for scalable video transmission over cellular networks, and 3) power-minimized joint power control and resource allocation for video communications over wireless channels. **Chapter 3** examines two resource allocation problems, which are: 1) distributed throughput maximization for scalable P2P VoD systems, and 2) streaming capacity for P2P VoD systems. In the throughput maximization problem, the distributed algorithms are presented to maximize the throughput for buffer-forwarding P2P VoD systems and hybrid-forwarding P2P VoD systems, respectively. In the streaming capacity problem, intra-channel and cross-channel resources are optimized to improve the streaming capacity for P2P VoD systems. **Chapter 4** presents an optimal prefetching framework, based on the segment access information, to reduce the seeking delay in P2P VoD applications. The segment access information is gathered with the hybrid sketches, which greatly reduces the space and time complexity. This chapter also gives an optimal substream allocation scheme in layered P2P VoD applications to improve the video quality. **Chapter 5** examines the distributed optimization techniques for unicast and multicast video streaming over wireless ad hoc networks. For unicast video streaming over wireless ad hoc networks, the expected distortion is minimized by jointly optimizing both the source rate and the routing scheme. For multicast video streaming over wireless ad hoc networks, a cross-layer optimization scheme is presented to jointly optimize the source rate, the routing scheme, and the power allocation for video streaming. **Chapter 6** examines the network lifetime maximization problem in wireless visual sensor networks. A distributed algorithm is developed to maximize the network lifetime by jointly optimizing the source rates, the encoding powers, and the routing scheme.

The target audience of the book includes researchers, educators, students, and engineers. The book can serve as a reference book in the undergraduate or graduate courses on multimedia communications and multimedia networking. It can also be used as a reference in research of multimedia communications and design of multimedia systems.

Yifeng He, Ling Guan, and Wenwu Zhu

1

Introduction

CONTENTS

1.1 Introduction

Recent years have witnessed a dramatic growth in multimedia applications, due to the tremendous advancement in several emerging technologies in digital media, communications, and networking. Multimedia data in a variety of popular formats such as image, animation, audio, and video have become the mainstream contents on the Internet and are being accessed by various mobile devices through wireless links. The diverse deployment of multimedia applications has been driving the field of multimedia communications and networking into a new era of true ubiquitous media consumption at any time and from anywhere.

Among all types of multimedia, digital video has been considered the dominating media form because of its volumetric spatial-temporal presentation nature. Over the past few years, digital video has already become the main traffic payload for Internet and major wireless networks. YouTube [1], a successful Internet video service, has been attracting a huge amount of users. According to the YouTube statistics [2], over 3 billion videos were viewed a day in 2010, more than 13 million hours of video were uploaded during 2010 and 48 hours of video are uploaded every minute. Annual global IP traffic will reach the zettabyte threshold (966 exabytes or nearly 1 zettabyte) by the end of 2015 [3]. Internet video was 40 percent of consumer Internet traffic in 2011, and will reach 61 percent by the end of 2015, not including the amount of video exchanged through Peer-to-Peer (P2P) file sharing. The sum of all forms of video (TV, Video on Demand (VoD), Internet, and P2P) will continue to be approximately 90 percent of global consumer traffic by 2015. The real-time video such as live video, ambient video, and video calling are taking an ever greater share of video traffic [3]. Globally, mobile data traffic will increase 26 times between 2010 and 2015 [4]. Mobile data traffic will grow at a Compound Annual Growth Rate (CAGR) of 92 percent between 2010 and 2015, reaching 6.3 exabytes per month by 2015. Because mobile video content has much higher bit rates than other mobile content types, mobile video will generate much of the mobile traffic growth through 2015. Mobile video traffic accounted for 52.8 percent of traffic in 2011 [4].

Many multimedia applications involve real-time video transmissions over distributed systems, in which there is no centralized controller. Examples of such distributed systems are Peer-to-Peer (P2P) networks, wireless ad hoc networks and Wireless Sensor Networks (WSNs). Though each type of distributed system has its own features, they share two common characteristics as follows. 1) Each node only knows about its neighbors, and does not have global knowledge. 2) There is typically no centralized controller, who can coordinate the behaviors of all the nodes. Therefore, a centralized algorithm is not practical for distributed systems. Instead, distributed algorithms are desired.

Since the appearance of Napster [5], a P2P file sharing service, in early 1999, P2P networks have experienced tremendous growth. In 2003, P2P became the most popular Web application. At the end of 2004, P2P traffic represented over 60% of the total Internet traffic, dwarfing Web browsing [6]. This rapid success was fueled by file transfer networks which allow users to swap the blocks of a file, despite the large time often necessary to complete a download. It is expected to continue at a fast pace, as new compelling P2P applications are developed. VoD has become an extremely popular service on the Internet. For example, YouTube [1] provides its services via servers. The deployment cost is expensive. It is appealing to apply P2P technology to VoD applications as P2P does not require any dedicated infrastructure and is self-scaling as the resources of the network increase with the number of users. In P2P VoD, a video is transported to a large number of asynchronous viewers by taking advantage of the uplink capability of the viewers to forward data. The major difference from P2P file sharing applications is that this should happen in real-time, to provide all connected users with a TV-like viewing experience.

Wireless ad hoc networks are multi-hop wireless networks without a pre-installed infrastructure. They can be deployed quickly at conventions, disaster recovery areas, and battlefields. When deployed, mobile nodes cooperate with each other to find routes and relay packets for each other. There is a compelling need for video streaming over wireless ad hoc networks. For example, a group of visitors in a museum would like to share their captured video in real-time. They can set up a wireless ad hoc network using their Personal Digital Assistants (PDAs), and then multicast the video to each group member.

A WSN is a system consisting of geographically distributed sensor nodes that communicate with each other over wireless channels. Without the need for a communication infrastructure, the WSN is self-organized and highly dynamic, with each node sensing and forwarding the data [1]. A Wireless Visual Sensor Network (WVSN) is a special kind of WSN, in that each sensor is equipped with video capture and processing components. A WVSN captures digital visual information about target events or situations and delivers the video data to a Remote Control Unit (RCU) for further information analysis and decision making. Because of its unique features of rapid deployment, flexibility, low maintenance cost, and robustness, WVSNs have been used in a wide range of important applications, including security monitoring, emergence response, environmental tracking, and health monitoring.

1.2 Technical Challenges in Distributed Video Communications

The Internet is a best-effort network, which cannot provide bandwidth, loss and delay guarantees. Wireless communications typically suffer from interference and channel fading. Despite the recent advances on video streaming over distributed systems, many challenges have not yet been well addressed.

In P2P streaming applications, high video quality and low startup delay are two major goals. However, it is difficult to achieve these two goals. First, the access bandwidth of the peers is often limited, especially the upload bandwidth. Second, the peers may leave the network at any time, which creates a highly unreliable and dynamic network fabric. Third, each user requests the video at a different time, and the video content is delivered from the heterogenous peers. Fourth, it has been observed that users frequently seek a different position rather than watching sequentially in VoD applications, which places a greater challenge on playback continuity.

In wireless ad hoc networks, a wireless link usually has a high transmission error rate because of shadowing, fading, and interferences from other transmitting users. An end-to-end path in ad hoc networks has an even higher error rate since it is the concatenation of multiple wireless links. Moreover, user mobility makes the network topology frequently change. An end-to-end route may only exist for a short period of time. The frequent link failures and route changes cause packet losses, thus degrading the received video quality.

In wireless visual sensor networks, each video sensor operates under a set of unique resource constraints, including limited energy supply, limited on-board computational capability, and low transmission bandwidth. In conventional wireless sensor networks, the power for signal processing at each sensor is very small. In contrast, the video sensor in WVSNs compresses the video before transmission. The compression takes a large amount of power, and raises a greater challenge for maintaining a long network lifetime.

1.3   Recent Advances in Optimal Resource Allocation for Distributed Video Communications

Optimal resource allocation provides an efficient solution to the problem for video communications over distributed systems. The problem in the distributed systems can be formulated into a resource allocation problem with an objective to maximize (or minimize) a performance metric, subject to the resource constraints at each node. The resource constraints include the flow conservation, the limitation of upload and download capacities, the limitation of buffer and storage capacities, the limitation of power supply, and the application-layer requirement (for example, the distortion requirement). Since there is no centralized controller in the distributed systems, a distributed algorithm is the desired solution in terms of the scalability and the communication overhead.

In this section, we provide a review of recent advances on optimal resource allocation for video communications over some major distributed systems including P2P streaming systems, wireless ad hoc networks, and wireless visual sensor networks.

1.3.1   Optimal Resource Allocation in P2P Streaming Systems

P2P streaming is one of the major applications on P2P overlay networks. In P2P streaming systems, each peer contributes its resources (e.g., bandwidth, buffer, and storage) to the community. Meanwhile, it retrieves the requested streams from other peers. The collaboration among peers enables each user to receive a high video quality. Depending on the applications, P2P streaming systems can be classified into P2P live streaming systems and P2P VoD streaming systems. In a P2P live streaming system, all of the users watch the same segment of the video at the same time, such that they can share the content with each other. In a P2P VoD streaming system, each user requests the video at a different time, therefore a peer can only request the content from the peer with an earlier playback progress.

1.3.1.1   P2P Live Streaming Systems

The optimization of the scheduling problem in data-driven P2P live streaming systems is studied in [8]. In data-driven P2P live streaming systems, the video is divided into blocks. There is an exchanging window at each peer containing the blocks that the peer is requesting. Every peer periodically notifies each of its neighbors the availability of the blocks. Then peers will request their absent blocks from neighbors. Different blocks have different priority. For instance, the blocks that have fewer suppliers should be requested preemptively such that they can be spread more quickly. Two factors have been considered in the priority definition: rarity factor and emergency factor. The rarity factor is considered first to guarantee the rarest block should be requested in priority, while the emergency factor is used to reduce the probability that the requested blocks miss the playback deadline. The objective of the optimal scheduling problem in [8] is to maximize the sum of priorities of all requested blocks in the overlay under the bandwidth constraints. To solve the optimal scheduling problem, a heuristic algorithm is developed, which is fully distributed and asynchronous with only local information exchange.

The end-to-end latency in P2P live streaming systems is investigated in [9], in which the optimization problem is formulated to minimize the average end-to-end streaming latency, subject to the constraints of the peer upload and download bandwidth. A distributed solution to the optimization problem is designed using Lagrangian decomposition and subgradient method. In the distributed algorithm, each peer carries out distributed steps with only local information, and such distributed execution achieves the global optimal objective.

In [10], Setton et al. propose that distortion-optimized retransmission requests are issued by receiving peers in a tree-based P2P live system to recover the most important missing packets while limiting the induced congestion. After detecting a parent disconnection, a peer can determine a list of missing packets and iteratively select the most important ones to request. This choice should depend on the time at which packets are due, and on the contribution of each packet to the overall video quality.

In the P2P application-layer overlay, multiple content distribution or media streaming sessions are expected to be running concurrently. It is important to provide differentiated services among the sessions. For example, a live streaming session should be handled with a higher priority than a content distribution session of bulk data. The problem of service differentiation across P2P communication sessions is examined in [11]. In order to select better paths and guarantee the rate for high-priority sessions, a utility function at link l for session s is given by $U_l^s = C_s \log(1 + Q_l x_l^s)$, where Q_l denotes the quality of link l, C_s is the priority of session s, and x_l^s denotes the bandwidth allocation to session s at link l. The optimization problem of service differentiation is formulated to maximize the summation of the utilities among all the sessions with the constraints of the heterogeneous upload and download capacities at each peer. A fully distributed algorithm is developed to solve the optimization problem using the subgradient method and Lagrangian relaxation.

Due to bandwidth constraint, most of the current P2P live streaming systems provide the video at a low bit rate. Users would like to watch the video at a higher quality. However, the P2P network may not be able to deliver the video at a high rate. Therefore, what is the upper bound of the streaming rate in a P2P system becomes an attractive topic. In [32], Sengupta et al. define the *streaming capacity* as the maximum supported streaming rate that can be received by every receiver, and compute the streaming capacity in a multi-tree P2P live streaming system. The streaming capacity problem is formulated to maximize the streaming rate r, given by $r = \sum_{t \in \mathbf{T}} y_t$ where $\mathbf{T}$ is the set of all allowed trees for the live session, and y_t is the rate of the sub-stream supported by tree t. The constraint in the streaming capacity problem is the upload capacity at each peer. An iterative combinatorial algorithm is designed to solve the streaming capacity problem approximately.

Most P2P live video systems offer a large number of channels, with users switching frequently among the channels. Wu et al. study the performance of multichannel P2P live video systems by using infinite-server queueing network models [43][14]. In multi-channel P2P streaming systems, optimal utilization of cross-channel resources can improve the system performance. In [15], a View-Upload Decoupling (VUD) scheme is proposed to decouple peer downloading from uploading, bringing stability to multichannel systems and enabling cross-channel resource sharing. In [16], Wang et al. formulate linear programming problems to maximize the sum of the bandwidth satisfaction ratios of all channels for three bandwidth allocation schemes, namely the Naive Bandwidth allocation Approach (NBA), the Passive Channel-aware bandwidth allocation Approach (PCA) and the Active Channel-aware bandwidth allocation Approach (ACA), respectively.

1.3.1.2   P2P VoD Streaming Systems

In P2P VoD streaming systems, the peers watching the same video can be organized into an overlay based on the playback progress, such that the peer with an earlier playback time can supply streams to the peer with a later playback time. The min-cost flow routing problem in P2P VoD systems is studied in [9]. The flow routing problem is formulated as a linear program with an objective of minimizing the aggregated link cost, subject to the inequality constraints of the peer upload and download capacities and the equality constraint that each peer has the same playback rate. A distributed auction algorithm is proposed to solve the min-cost flow routing problem.

Most of the existing P2P VoD systems only adopt single-layer video coding [9, 11]. If the source rate is high, the peers with a limited or low download bandwidth may not be able to accommodate it. On the other hand, if the source rate is too low, the peers with a higher download bandwidth may underutilize their download bandwidth. Therefore, a scalable source coding is a good solution for P2P VoD applications with heterogeneous bandwidth. In a scalable P2P VoD system, one of the goals is to maximize the aggregate throughput over all the peers. In [19], the throughput maximization problem in the scalable P2P VoD system is formulated to maximize the aggregate throughput by optimally allocating the link rates, subject to the *source rate constraint* representing that the received rate at any peer is no larger than the maximal source rate, the *download bandwidth constraint*, the *upload bandwidth constraint*, and the *link-forwarding constraint* representing that each outgoing link from peer i carries a rate no larger than the total incoming rate into the peer. A distributed algorithm is developed to solve the throughput maximization problem using Lagrangian decomposition and subgradient method.

Depending on the forwarding approach, the existing P2P VoD systems can be classified into two categories: buffer-forwarding architecture [9, 10] and storage-forwarding architecture [25, 11]. In buffer-forwarding architectures, each peer buffers the recently received content, and forwards it to the child peers. In storage-forwarding architectures, the video content is distributed over the storage of peers. When a peer wants to watch a video, it first looks for the serving peers who are storing the content, and then requests the content from them in parallel. In order to fully utilize the resources in the P2P VoD systems, a hybrid architecture integrating both the buffer-forwarding approach and storage-forwarding approach is proposed [22, 23]. The throughput maximization problem in the hybrid P2P VoD architecture is formulated, and a distributed algorithm is designed to maximize the throughput by optimizing the link rates.

Unlike the users in P2P live streaming systems who watch the broadcast video passively, the users in P2P VoD streaming systems may seek to any position that he or she is interested in, as demonstrated in [3]. The behaviors of random seek place a great challenge to the playback continuity. Zheng et al. propose a prefetching scheme to improve the playback continuity [3]. The user seeking pattern is obtained from the previous seeking statistics. Based on the seeking pattern, the segments that will be prefetched are determined optimally to minimize the expected seeking distance, the deviation between the desired seeking position and the scheduled position. The Lloyd algorithm is used to solve the prefetching optimization problem.

The work in [3] represents the seeking pattern using one-dimensional Probability Density Function (PDF) $P(y)$ where y is the destination segment of a seek. In order to capture the seeking behaviors in a more accurate way, the work in [25] uses two-dimensional PDF, denoted as $P(x, y)$, representing the probability that a user performs a seek from the start segment x to the destination segment y. In [26], the concept of *guided seek* is introduced. With the guidance of the two-dimensional PDF, users can perform efficient seeks to the desired positions. The guidance can be obtained from collective seeking statistics of other peers who have watched the same title in the previous and/or concurrent sessions. *Hybrid sketches* are designed to capture the seeking statistics at significantly reduced space and time complexity. Furthermore, an optimal prefetching scheme and an optimal cache replacement scheme are proposed to minimize the expected seeking delay by optimally determining the segments to be prefetched.

In P2P VoD streaming systems, a peer can only request the desired segments from the peer with an earlier playback progress. Therefore, the streaming capacity in P2P VoD streaming systems is different from that in P2P live streaming systems. The streaming capacity in a P2P VoD streaming system is formulated to maximize the streaming rate that can be received by every peer, subject to the limitations of upload and download capacities at each peer [27][37]. In [29], helpers are introduced in the P2P VoD system and the helper resources are optimized to improve the streaming capacity. The streaming capacity for multi-channel P2P VoD systems is studied in [46], in which the cross-channel resource sharing schemes are proposed to maximize the average streaming capacity.

1.3.2   Optimal Resource Allocation for Video Streaming over Wireless Ad Hoc Networks

Depending on the number of simultaneous receivers, video streaming over wireless ad hoc networks can be classified into two classes: unicast and multi-cast [1].

In unicast video streaming, path diversity is very attractive since it provides an effective means to combat transmission errors in wireless ad hoc networks. In [32], a path selection scheme is proposed for multi-path video streaming over wireless ad hoc networks. The optimization problem is formulated to minimize the concurrent Packet Drop Probability (PDP) by selecting two optimal paths.

The approach of selecting paths by minimizing packet loss [32] does not guarantee the minimization of the expected video distortion. A distortion-minimized scheme is proposed in [33] for unicast video streaming over wireless ad hoc networks. The received video distortion consists of the encoding distortion and the transmission distortion, which is given by $D_{dec} = D_{enc} + D_{tran}$. The encoding distortion D_{enc} is introduced by the quantization at the encoder, and it can be calculated by $D_{enc} = D_0 + \theta/(R - R_0)$ where R is the rate of the video, and (D_0, θ, R_0) are the parameters relative to video encoding. The transmission distortion is caused by the packet loss, and it can be calculated by $D_{tran} = \kappa(P_r + (1 - P_r)e^{-(C'-R)T/L'})$ where P_r is the random packet loss due to transmission errors, and (κ, C', T, L') are positive parameters which are discussed in [33]. At lower rates, reconstructed video quality is limited by coarse quantization, whereas at high rates, the video stream will cause more network congestion and therefore will lead to longer packet delays. These, in turn, translate into higher loss rates, hence reduced video quality. Therefore, the optimal video rate can be obtained by $R^* = \arg\min D_{dec}$.

The combination of Multiple Description Coding (MDC) and multi-path transport has shown the superiority in terms of error resilience [6]. Optimization of the routing scheme for a single unicast video session in wireless ad hoc networks is examined in [14]. The video is encoded into two MDC descriptions. The optimization problem is formulated to minimize the expected video distortion by finding the optimal path for each description. A Genetic Algorithm (GA) based solution is provided to solve the optimization problem.

The work in [32, 33, 14] studies the unicast video streaming from a single sender to a single receiver. The optimization of a multi-path routing scheme for a video streaming session from two senders to a single receiver is presented in [36]. The optimization problem is formulated to minimize the expected distortion of the MDC video by selecting the optimal sender and the optimal path from the sender to the receiver for each MDC description.

Joint optimization of the source rate and the routing scheme for unicast video streaming is studied in [37]. The received video distortion is given by $D_{dec} = D_0 + \theta_0/((s_r - \sum_{l \in \mathbf{L}} p_l x_l) + \phi_0)$ where s_r is the source rate, x_l is the link rate at link l, p_l is the packet loss rate at link l, and (D_0, θ_0, ϕ_0) are the predetermined parameters. The optimization problem is stated as follows. Given a wireless ad hoc network and a unicast session from a sender to a receiver, to minimize the received video distortion D_{dec} by optimally determining the source rate s_r and the link rate x_l at link l, subject to the network flow constraint and the link capacity constraint. A distributed algorithm using dual decomposition is developed to solve the optimization problem.

Different from the single video session, multiple concurrent video sessions in a wireless ad hoc network have to compete for limited network resources. Such interactions make the performance of an individual flow couple with that of other flows. Joint optimization of the source rate and the routing scheme for multiple concurrent video sessions in wireless ad hoc networks is investigated in [38, 23]. In [38], the optimization problem is formulated to minimize the sum of the average distortions of all concurrent video sessions by jointly optimizing the source rate and the routing path for each video session. A greedy heuristic algorithm is proposed to find the near-optimal solution to the optimization problem. In [23], the objective of the optimization problem is to minimize the Lagrangian sum of the total video distortion and the overall network congestion. A distributed algorithm is proposed to solve the optimization problem.

Multicast video streaming over wireless ad hoc networks is bandwidth-efficient compared to multiple unicast sessions. Multiple-tree routing algorithms are proposed to explore the path diversity for each receiver. Two typical multiple-tree video multicast schemes in wireless ad hoc networks are presented in [26, 25], respectively.

In [26], Zakhor et al. propose a multiple-tree construction protocol that builds two nearly disjoint trees simultaneously in a distributed manner. However, the trees are built based on the network-layer metrics, and the application performance has not yet been optimized.

In [25], two multicast trees are constructed to deliver two MDC descriptions. Each description is layered, encoded to meet the heterogeneous capacity of the receivers. The optimization problem is formulated to minimize the total video distortion of all receivers by constructing two optimal trees. A GA-based heuristic is proposed to solve the optimization problem.

A joint optimization of the source rate and the routing scheme for multi-cast video streaming over wireless ad hoc networks is presented in [42]. A joint optimization of the source rate, the routing scheme, and the power allocation for multicast video streaming is studied in [43]. In [43], a prioritized coding scheme, a combination of the layered source coding and the network coding [4], is employed to enable the heterogeneous receivers to reconstruct the video at different quality levels. The network coding eliminates the delivery redundancy. Therefore, a larger throughput at a receiver can lead to a smaller distortion. The optimization problem is stated as follows. Given a wireless ad hoc network and a multicast session from a sender to multiple receivers, to maximize the aggregate throughput by optimally determining the source rate, the routing scheme and the power allocation, subject to the constraint of flow conservation, the link capacity constraint, and the power constraint. A distributed algorithm using hierarchical dual decompositions is developed to solve the optimization problem.

1.3.3 Optimal Resource Allocation for Wireless Visual Sensor Networks

Sensor nodes are typically battery powered, and battery replacement is infrequent or even impossible in many sensing applications. Hence, a tremendous amount of research efforts in wireless sensor networks has been focused on energy conservation. One aspect of this research is to maximize the network lifetime, which is typically defined as the time from the start of the network until the death of the first node [4, 7]. In [4], the network lifetime maximization problem is formulated into a convex optimization problem [19]. A distributed algorithm using Lagrangian decomposition is developed to maximize the network lifetime by optimizing the routing scheme.

The performance of wireless sensor network applications is typically a function of the amount of data collected by the individual sensors. There is an inherent trade-off in simultaneously maximizing the network lifetime and the application performance. Such tradeoff is investigated in [7, 6]. In [7], the application performance is characterized by a network utility function, which is strictly concave and increasing with respect to the data rate. The network lifetime is the minimum of the node lifetime, which is characterized by a lifetime-penalty function for each node. The optimization problem is formulated to maximize the Lagrangian sum of the network utility and the network lifetime, by jointly optimizing the source rates and the routing scheme.

The algorithms [4] that maximize the network lifetime in conventional wireless ad hoc networks cannot be applied directly to the wireless visual sensor networks, since they omit the encoding power consumption at the sensor nodes. The network lifetime maximization for wireless visual sensor networks is studied in [49, 50]. In [50], the relationship among the encoding power, the source rate, and the encoding distortion is characterized by $d_{sh} = \sigma^2 e^{-\gamma \cdot R_h \cdot P_{sh}^{2/3}}$ [16], where d_{sh} is the encoding distortion of the video encoded at sensor h, R_h is the source rate generated at sensor h, P_{sh} is the encoding power consumption at sensor h, σ^2 is the average input variance, and γ is the encoding efficiency coefficient. The problem of maximum network lifetime is stated as follows. Given the topology of a static WVSN and the initial energy at each node, to maximize the network lifetime by jointly optimizing the source rate and the encoding power at each video sensor, and the link rates for each session, subject to the constraint of flow conservation and the requirement of the collected video quality. A fully distributed algorithm using the properties of Lagrangian duality is developed to solve the network lifetime maximization problem.

The ultimate goal of the WVSN is to utilize its limited resources to collect as much visual information as possible. A metric, called *Accumulative Visual Information* (AVI), is introduced in [8] to measure the amount of visual information collected by the video sensor. The AVI is a function of the bit rate R and the encoding power P, which is given by $I = f(R, P)$. The optimal resource allocation problem in WVSN can be formulated to maximize the AVI by optimizing the allocation of the bit rate and encoding power for each frame of the video.

The power consumption P_0 at a video sensor mainly consists of the encoding power P_s and the transmission power P_t [2], which is given by $P_0 = P_s + P_t$. If the encoding power P_s is decreased, the distortion d of the video is increased, i.e., $P_s \downarrow \Rightarrow d \uparrow$. On the other hand, since the total power consumption P_0 is fixed, the transmission power P_t will be decreased if the encoding power P_s is increased. This implies that fewer bits can be transmitted because the transmission energy is proportional to the number of bits to be transmitted. Therefore, $P_s \uparrow \Rightarrow d \uparrow$. In other words, when the encoding power P_s goes too low or too high, the distortion d will become large. The distortion can be minimized by optimally allocating the encoding power P_s and the transmission power P_t [2].

1.4 Overview of the Book

This book examines the techniques for optimal resource allocation for distributed video and multimedia communications over Internet, wireless cellular networks, wireless ad hoc networks, and wireless sensor networks.

In **Chapter 2**, we present the resource allocation techniques for scalable video communications over the Internet or wireless networks. Specifically, we examine four topics: 1) network-adaptive resource allocation for scalable video streaming over the Internet, 2) Quality of Service (QoS) adaptive resource allocation for scalable video transmission over cellular networks, and 3) power-minimized joint power control and resource allocation for video communications over wireless channels.

In **Chapter 3**, we examine two resource allocation problems, which are: 1) distributed throughput maximization for scalable P2P VoD systems, and 2) streaming capacity for P2P VoD systems. In the throughput maximization problem, we present distributed algorithms to maximize the throughput for buffer-forwarding P2P VoD systems and hybrid-forwarding P2P VoD systems, respectively. In the streaming capacity problem, we optimize both intra-channel resource allocation and cross-channel resource allocation to improve the streaming capacity for P2P VoD systems.

In **Chapter 4**, we present an optimal prefetching framework to reduce the seeking delay in P2P VoD applications. We design the hybrid sketches to represent the seeking statistics, thus greatly reducing the space and time complexity. Based on the segment access information, we develope an optimal prefetching scheme and an optimal cache replacement scheme to minimize the expected seeking delay at every viewing position. We also present an optimal substream allocation scheme in layered P2P VoD applications to improve the video quality.

In **Chapter 5**, we examine the distributed optimization techniques for unicast and multicast video streaming over wireless ad hoc networks. For unicast video streaming over wireless ad hoc networks, we minimize the expected distortion by jointly optimizing both the source rate and the routing scheme. For multicast video streaming over wireless ad hoc networks, we examine video streaming under different communication models including Frequency Division Multiple Access (FDMA) and Code Division Multiple Access (CDMA). In video multicasting over FDMA wireless ad hoc networks, we optimize both the source rate and the routing scheme to improve the video quality. In video multicasting over CDMA wireless ad hoc networks, we present a cross-layer optimization scheme to jointly optimize the source rate, the routing scheme, and the power allocation for video streaming.

In **Chapter 6**, we examine the network lifetime maximization problem in wireless visual sensor networks. We first investigate the achievable maximum network lifetime in WVSN without transmission errors. We then further examine the maximum network lifetime considering transmission errors. We investigate the error remedy techniques in both large-delay WVSN applications and small-delay WVSN applications, and study their impacts on maximum network lifetime. We derive distributed algorithms by using dual decomposition to maximize the network lifetime by jointly optimizing the source rates, the encoding powers, and the routing scheme.

Bibliography

[1] http://www.youtube.com/

[2] http://www.youtube.com/t/press_statistics

[3] Cisco White Paper, "Entering the Zettabyte Era," Jun. 2011. http://www.cisco.com

[4] Cisco White Paper, "Cisco Visual Networking Index: Global Mobile Data Traffic Forecast Update, 2010 C2015," Feb. 2011. http://www.cisco.com

[5] http://www.napster.com

[6] M. Meeker, "The State of the Internet," in *Web 2.0 conference,* Nov. 2006.

[7] I. Akyildiz, W. Su, Y. Sankarasubramaniam, and E. Cayirci, "A survey on sensor networks," *IEEE Communication Magazine,* no. 8, pp. 102-114, Aug. 2002.

[8] M. Zhang, Y. Xiong, Q. Zhang, and S. Yang, "On the Optimal Scheduling for Media Streaming in Data-driven Overlay Networks," in *Proc. of IEEE GLOBECOM,* pp. 1-5, Nov. 2006.

[9] C. Wu and B. Li, "Optimal Peer Selection for Minimum-Delay Peer-to-Peer Streaming with Rateless Codes," in *Proc. of ACM MM,* pp. 69-78, Nov. 2005.

[10] E. Setton, J. Noh, and B. Girod, "Rate-Distortion Optimized Video Peer-to-Peer Multicast Streaming," in *Proc. of ACM MM,* pp. 39-45, Nov. 2005.

[11] C. Wu and B. Li, "Diverse: Application-Layer Service Differentiation in Peer-to-Peer Communications," *IEEE Journal on Selected Areas in Communications,* vol. 25, no. 1, pp. 222-234, Jan. 2007.

[12] S. Sengupta, S. Liu, M. Chen, M. Chiang, J. Li, and P. A. Chou, "Streaming Capacity in Peer-to-Peer Networks with Topology Constraints," *Microsoft Research Technical Report,* 2008.

[13] D. Wu, Y. Liu, and K.W. Ross, "Queuing network models for multichannel P2P live streaming systems," in *Proc. of IEEE INFOCOM,* pp. 73-81, Apr. 2009.

[14] D. Wu, Y. Liu, and K.W. Ross, "Modeling and Analysis of Multichannel P2P Live Video Systems," *IEEE/ACM Transactions on Networking, VOL. 18, NO. 4,,* vol. 18, no. 4, pp. 1248-1260, Aug. 2010.

[15] D. Wu, C. Liang, Y. Liu, and K.W. Ross, "View-Upload Decoupling: A Redesign of Multi-Channel P2P Video Systems," in *Proc. of IEEE INFO-COM*, pp. 2726-2730, Apr. 2009.

[16] M. Wang, L. Xu and B. Ramamurthy "Linear Programming Models For Multi-Channel P2P Streaming Systems," in *Proc. of IEEE INFOCOM*, pp. 1-5, Mar. 2010.

[17] Z. Li and A. Mahanti, "A Progressive Flow Auction Approach for Low-Cost On-Demand P2P Media Streaming," in *Proc. of ACM QShine*, Aug. 2006.

[18] W. P. Yiu, X. Jin and S. H. Chan, "Distributed storage to support user interactivity in peer-to-peer video streaming," in *Proc. of IEEE ICC*, vol. 1, pp. 55-60, Jun. 2006.

[19] Y. He, I. Lee, and L. Guan, "Distributed rate allocation in p2p streaming," in *Proc. of ICME*, pp. 388-391, Jul. 2007.

[20] C. Huang, J. Li, and K. W. Ross, "Peer-Assisted VoD: Making Internet Video Distribution Cheap," in *Proc. of IPTPS*, Feb. 2007.

[21] Y. Shen, Z. Liu, S. S. Panwar, K. W. Ross, and Y. Wang, "Streaming Layered Encoded Video Using Peers," in *Proc. of ICME*, Jul. 2005.

[22] Y. He, I. Lee, and L. Guan, "Distributed throughput maximization in hybrid-forwarding P2P VoD applications," in *Proc. of ICASSP*, pp. 2165-2168, Apr. 2008.

[23] Y. He, I. Lee, and L. Guan, "Distributed throughput optimization in P2P VoD applications," *IEEE Transactions on Multimedia*, vol. 11, no. 3, pp. 509-522, Apr. 2009.

[24] C. Zheng, G. Shen, and S. Li, "Distributed Prefetching Scheme for Random Seek Support in Peer-to-Peer Streaming Applications," in *Proc. of ACM MM*, pp. 29-38, Nov. 2005.

[25] Y. He, G. Shen, Y. Xiong and L. Guan, "Probabilistic prefetching scheme for P2P VoD applications with frequent seeks," in *Proc. of ISCAS*, pp. 2054-2057, May 2007.

[26] Y. He, G. Shen, Y. Xiong and L. Guan "Optimal prefetching scheme in P2P VoD applications with guided seeks," *IEEE Transactions on Multimedia*, vol. 11, no. 1, pp. 138-151, Jan. 2009.

[27] Y. He and L. Guan, "Streaming capacity in P2P VoD systems," in *Proc. of ISCAS*, pp. 742-745, May 2009.

[28] Y. He and L. Guan, "Solving streaming capacity problems in P2P VoD systems," *IEEE Transactions on Circuits and Systems for Video Technology*, vol. 20, no. 11, pp. 1638 C1642, Nov. 2010.

[29] Y. He and L. Guan, "Improving the streaming capacity in P2P VoD systems with helpers," in *Proc. of ICME,* pp. 790-793, Jul. 2009.

[30] Y. He and L. Guan, "Streaming capacity in multi-channel P2P VoD systems," in *Proc. of ISCAS,* pp. 1819-1822, May 2010.

[31] W. Wei and A. Zakhor, "Multipath unicast and multicast video communication over wireless ad hoc networks," in *Proc. of IEEE BroadNets,* pp. 496-505, Oct. 2004.

[32] W. Wei and A. Zakhor, "Path Selection for Multi-Path Streaming in Wireless Ad Hoc Networks," in *Proc. of IEEE ICIP,* pp. 3045-3048, Oct. 2006.

[33] X. Zhu, E. Setton, and B. Girod, "Congestion-Distortion Optimized Video Transmission over Ad Hoc Networks," *Journal of Signal Processing: Image Communications,* no. 20, pp. 773-783, Sep. 2005.

[34] S. Mao, S. Lin, S. Panwar, Y. Wang, and E. Celebi, "Video transport over ad hoc networks: Multistream coding with multipath transport," *IEEE Journal on Selected Areas in Communications,* vol. 21, no. 10, pp. 1721-1737, Dec. 2003.

[35] S. Mao, Y. T. Hou, X. Cheng, H. D. Sherali, and S. F. Midkiff, "Multipath routing for multiple description video over wireless ad hoc networks," in *Proc. of IEEE INFOCOM,* pp. 740-750, Mar. 2005.

[36] S. Mao, X. Cheng, Y. T. Hou, H. D. Sherali, and J. H. Reed, "Joint Routing and Server Selection for Multiple Description Video Streaming in Ad Hoc Networks," in *Proc. of IEEE ICC,* vol. 5, pp. 2993-2999, May 2005.

[37] Y. He, I. Lee, and L. Guan, "Optimized multi-path routing using dual decomposition for wireless video streaming," in *Proc. of ISCAS,* pp. 977-980, May 2007.

[38] S. Mao, S. Kompella, Y. T. Hou, H. D. Sherali, and S. F. Midkiff, "Routing for Concurrent Video Sessions in Ad Hoc Networks," *IEEE Transactions on Vehicular Technology,* vol. 55, no. 1, pp. 317-327, Jan. 2006.

[39] X. Zhu, J. P. Singh, and B. Girod, "Joint routing and rate allocation for multiple video streams in ad hoc wireless networks," *Journal of Zhejiang University,* Science A, vol. 7, no. 5, pp. 727-736, May 2006.

[40] A. Zakhor and W. Wei, "Multiple Tree Video Multicast over Wireless Ad Hoc Networks," in *Proc. of ICIP,* pp. 1665-1668, Sep. 2006.

[41] S. Mao, X. Cheng, Y. T. Hou, and H. D. Sherali, "Multiple description video multicast in wireless ad hoc networks," in *Proc. of IEEE BROAD-NETS,* pp. 671-680, Oct. 2004.

[42] Y. He, I. Lee, and L. Guan, "Video multicast over wireless ad hoc networks using distributed optimization," in *Proc. of IEEE PCM*, pp. 296-305, Dec. 2007.

[43] Y. He, I. Lee, and L. Guan, "Optimized video multicasting over wireless ad hoc networks using distributed algorithm," *IEEE Transactions on Circuits and Systems for Video Technology*, vol. 19, no. 6, pp. 796-807, Jun. 2009.

[44] R. Ahlswede, N. Cai, S.-Y. R. Li and R. W. Yeung, "Network information flow," *IEEE Transactions on Information Theory*, vol. 46, pp. 1204-1216, Jul. 2000.

[45] R. Madan, S. Lall, "Distributed algorithms for maximum lifetime routing in wireless sensor networks," *IEEE Transactions on Wireless Communications*, vol. 5, no. 8, pp. 2185-2193, Aug. 2006.

[46] H. Nama, M. Chiang, and N. Mandayam, "Utility-lifetime trade-off in self-regulating wireless sensor networks: A cross-layer design approach," in *Proc. of IEEE ICC*, vol. 8, pp. 3511-3516, Jun. 2006.

[47] S. Boyd and L. Vandenberghe, *Convex Optimization*, Cambridge University Press, 2004.

[48] J. Zhu, K. Hung, B. Bensaou, and F. Abdesselam, "Tradeoff between network lifetime and fair rate allocation in wireless sensor networks with multi-path routing," in *Proc. of ACM MSWiM*, pp. 301-308, Oct. 2006.

[49] Y. He, I. Lee, and L. Guan, "Network lifetime maximization in wireless visual sensor networks using a distributed algorithm," in *Proc. of ICME*, pp. 2174-2177, Jul. 2007.

[50] Y. He, I. Lee, and L. Guan, "Distributed algorithms for network lifetime maximization in wireless visual sensor networks," *IEEE Transactions on Circuits and Systems for Video Technology*, vol. 19, no. 5, pp. 704-718, May 2009.

[51] Z. He, Y. Liang, L. Chen, I. Ahmad, and D.Wu, "Power-rate-distortion analysis for wireless video communication under energy constraint," *IEEE Transactions on Circuits and Systems for Video Technology*, vol. 15, no. 5, pp. 645-658, May 2005.

[52] Z. He and D. Wu, "Accumulative visual information in wireless video sensor network: definition and analysis," in *Proc. of IEEE ICC*, vol. 2, pp. 1205-1208, May 2005.

[53] Z. He and D. Wu, "Resource allocation and performance analysis of wireless video sensors," *IEEE Transactions on Circuits and Systems for Video Technology*, vol. 16, no. 5, pp. 590-599, May 2006.

2

Optimized Resource Allocation for Scalable Video Communications

CONTENTS

In video communications over networks, there is a tradeoff between video compression and video transmission. In other words, there is a tradeoff between the bitrate allocated for video source coding and the bit rate allocated for channel coding, under a given bandwidth capacity constraint. Scalable Video Coding (SVC) is a promising source coding technique which can adapt the video quality to the available network bandwidth. SVC is the Annex G extension of the H.264 video compression standard. It is composed of one Base Layer (BL) and one or multiple Enhancement Layers (ELs). Forward Error Correction (FEC) is a channel coding technique, which has been widely used in real-time video applications to satisfy the strict delay requirements.

In this chapter, we will present the techniques for optimal resource allocation for scalable video communications over the Internet or wireless networks. Specifically, we will examine the following four topics: 1) network-adaptive resource allocation for scalable video streaming over the Internet, 2) Quality of

Service (QoS)-adaptive resource allocation for scalable video transmission over cellular networks, and 3) power-minimized joint power control and resource allocation for video communications over wireless channels.

2.1   Network-Adaptive Resource Allocation for Scalable Video Streaming over the Internet

Audiovisual streaming over the Internet has become very popular today. Technically, delivery of streaming media over the Internet with Quality of Experience (QoE) has many challenges. On one hand, the current Internet only provides best-effort service and it does not provide QoS guarantees or provision for multimedia services. Specifically, network conditions and characteristics, such as bandwidth, packet loss ratio, delay, and delay jitter, vary from time to time. On the other hand, media encoders generally do not take the network conditions into account. In general, different kinds of media have different characteristics. Real-time media such as video or audio is delay sensitive but capable of tolerating a certain degree of errors. Non-real-time media, such as Web data, is less delay sensitive but requires reliable transmission. Consequently, different types of media may have different quality impairments under the same network conditions. Therefore, designing a high-quality media streaming system that can cope with varying Internet conditions becomes important.

In the literature, several schemes have been developed for QoS management, including resource reservation, priority mechanism, and application control. Resource Reservation Protocol (RSVP) is the most straightforward approach [3]. However, RSVP requires all routers to have QoS supports. In addition, it may tend to over-allocate resources for QoS guarantee, thus leading to network under-utilization. In priority-based mechanisms, different data packets or streams are labeled with different priorities and treated differently in the network routers. This is also called Differentiated Service (DiffServ). However, the exact mechanism for setting the priority levels and mapping from the application priority levels to the router priority levels, the router mechanism for controlling these levels, and the performance gains for defining priority classes are under investigation [32, 33]. In the application control, the QoS is enforced by congestion control and transmission-rate adaptation [29, 16, 30]. The advantage is that there is almost no need to change the router or network itself. However, the main challenge is to design efficient congestion-and-flow control.

To efficiently transport media over the Internet, both real-time and non-real-time systems are expected to react to congestion by adapting their transmission rates and maintaining the inter-protocol fairness. Since a dominant portion of today's Internet traffic is TCP-based, it is very important for multimedia streams be "TCP-friendly," by which we mean a media flow generates a similar throughput as a typical TCP flow along the same path under the same conditions with lower latency. There are two existing groups of TCP-friendly flow-control protocols for multimedia streaming applications: sender-based rate adjustment and model-based flow control. Sender-based rate adjustment [29, 16, 4] performs additive increase and multiplicative decrease (AIMD) rate control in the sender as in TCP. The transmission rate is increased in a step-like fashion in the absence of packet loss and reduced multiplicatively when congestion is detected. This approach usually requires the receiver to send an acknowledgement for every received packet to detect congestion indications, such as packet loss and timeouts. The drawbacks of this approach are: 1) network congestion could severely degrade the performance since frequent feedback packets are needed for flow control; 2) the time-varying network status cannot be reflected since the control scheme is independent of packet loss ratio, bandwidth variation, and adjusting interval. Model-based flow control [24, 36, 34], on the other hand, uses a stochastic TCP model [23], which represents the throughput of a TCP sender as a function of packet loss ratio and Round Trip Time (RTT). Since this protocol can run in the receiver, the congestion problem in the reverse path can be avoided. However, this approach also has its shortcomings. First, the available bandwidth may be over-estimated or under-estimated for high packet loss ratio. Second, the estimated packet loss ratio is not for the next time interval so as to affect the accuracy of the throughput calculation. Third, sending rate is re-assigned to meet the calculated bandwidth, and its fluctuation is not suitable for continuous media.

In some applications where audio, video, and data, or a set of visual elements are delivered simultaneously over the Internet, the media rates are usually aggregated. To make the aggregated bit rate equal to or less than the Internet available bandwidth, independent control for each media is usually employed by allocating a fixed rate to each media. However, this may lead to large variations in quality among different media and thus cause inefficient utilization of the Internet resource. Unlike independent control, joint control only needs to maintain a constant aggregate bit rate while allowing bit rate of each media to vary. Recent studies have shown that joint control is more efficient than independent control for multiple media coding [39, 37, 10]. However, none of these approaches takes the time-varying network conditions into account. Since different media may have different quality degradations under various network situations, it is intuitive to move bits from less active and lightly degraded media to more active and heavily degraded ones.

To address the above issues, we present a network-adaptive bit allocation scheme for a multi-layer scalable video codec. Specifically, we first present a new Multimedia Streaming TCP-friendly Protocol (MSTFP) to iteratively combine forward estimation of network condition with information feedback control to optimally track network status. The MSTFP is well suited for continuous media streaming since it integrates accurate throughput calculation with history-related rate adjustment. Then, we present a novel resource allocation scheme for multiple media streams to achieve end-to-end optimal quality according to the estimated network bandwidth and media rate-distortion functions.

In the following bitrate-allocation formulation, we use the Progressive Fine Granularity Scalable (PFGS) video codec as an example although our approach can be applied to any scalable codec such as MPEG Fine Granularity Scalable (FGS) [12] or H.264 SVC. We combine PFGS codec with network-adaptive Unequal Error Protection (UEP) across packets. We strongly protect the base layer of PFGS against packet loss so as to be decodable even if no enhancement layers are available by employing UEP based on Reed–Solomon (RS) FEC code.

The difficulty encountered in joint bit allocation between source and Internet channel is how to add FEC so that the decoder can recover the lost frames correctly. Obviously, it can be observed that under a given channel rate, the additional FEC packets reduce the available rate for source coding, thus resulting in a trade-off between source coding and FEC. In the bit rate-allocation formulation, we address the problem on the optimal bit rate allocation between the source and the FEC based on the Rate–Distortion (R–D) function such that the decoder can successfully recover the lost packets. Specifically, the optimal bit allocation is dynamically adjusted according to varying video characteristic and network condition. We formulate this problem as follows. Let $R(t)$ denote the network bandwidth available for transmission at time t. Let $R_S(t)$ and $R_{FEC}(t)$ denote PFGS source rate and the rate of FEC packets, respectively. Furthermore, let $D_S(t)$ and $D_{FEC}(t)$ represent PFGS source distortion and distortion from FEC packets, respectively. Then, this problem is to allocate the available bit rate at time t to minimize the end-to-end video distortion by optimizing $R_S(t)$ and $R_{FEC}(t)$ under the following constraint: $R_S(t) + R_{FEC}(t) \leq R(t)$. Mathematically, the problem is formulated as:

$$
\begin{aligned}
\text{Minimize} \quad & D = D_S(t) + D_{FEC}(t) \\
\text{subject to} \quad & R_S(t) + R_{FEC}(t) \leq R(t).
\end{aligned}
\tag{2.1}
$$

The block diagram of our bit allocation scheme for the PFGS source and UEP is illustrated in Figure 2.1. PFGS source coder encodes input video into two layers: one is the base layer (BL) that carries the most important information; the other is the enhancement layer (EL) that carries less important information. The EL bit stream can be truncated anywhere. These layers are packetized and protected against packet loss according to their importance and network status using different FEC. The channel estimation module adaptively updates the network status. On the receiver side, the channel decoder reconstructs packets for each layer and displays video after source decoding. To efficiently deliver video over the Internet, several error resilience mechanisms have been adopted in the video coder, such as error localization, data partition, error concealment, etc.

The idea of FEC across packets is to transmit additional packets that can

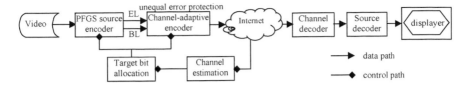

FIGURE 2.1
Block diagram of our network-adaptive bit allocation scheme for PFGS streaming with UEP [43]

be used in the receiver to reconstruct lost packets. Here the FEC scheme uses RS codes across packets. RS codes are perfectly suitable for error protection against packet loss, because they are the only known non-trivial maximum distance separable codes, i.e., there are no other existing codes that can reconstruct erased symbols from a smaller fraction of received code symbols [15]. An RS(n, k) code with length n and dimension k encodes k information symbols containing m bits per symbol into a codeword of n symbols. With the knowledge of error position, RS(n, k) can generally correct up to $t = n - k$ symbol errors.

To evaluate the performance of an RS(n, k) code, we need to know the probability that more than $n - k$ packets are lost. We can compute this probability if we know the probability of which m packets are lost within n packets.

As stated in [43], we use the 2-state Markov model to estimate network status. Such a model is determined by the distribution of error-free intervals (gap). Let gap length v be the event that after a lost packet, $v - 1$ packets are received and then another packet is lost. The gap density function $g(v)$ gives the probability of gap length v, i.e., $g(v) = \Pr\left(1^{v-1}0|0\right)$. The gap distribution function $G(v)$ is the probability of gap length greater than $v - 1$, i.e., $G(v) = \Pr\left(1^{v-1}|0\right)$. They can be derived as

$$g(v) = \begin{cases} 1 - p & for \ v = 1, \\ p(1-q)^{v-2}q & for \ v > 1, \end{cases} \qquad (2.2)$$

$$G(v) = \begin{cases} 1 & for \ v = 1, \\ p(1-q)^{v-2} & for \ v > 1. \end{cases} \qquad (2.3)$$

Let $R(m, n)$ be the probability of $m - 1$ packet losses within the next $n - 1$ packets followed by a lost packet. It can be calculated using recurrence as follows:

$$R(m, n) = \begin{cases} G(n) & for \ m = 1, \\ \sum_{v=1}^{n-m+1} g(v)R(m-1, n-v) & for \ 2 \le m \le n. \end{cases} \qquad (2.4)$$

Then, the probability of m lost packets within n packets is

$$P(m,n) = \sum_{v=1}^{n-m+1} P_B G(v) R(m, n-v+1) \quad for \ \ 1 \le m \le n, \qquad (2.5)$$

where P_B is the average of packet-loss probability.

Now, the probability that more than $n - k$ packets are lost within the n packets can be represented as $\sum_{m=n-k+1}^{n} P(m,n)$. This probability is the residual loss probability experienced by a video decoder after RS decoding, which can be used to design the overall system if how many losses are acceptable for a video decoder is known.

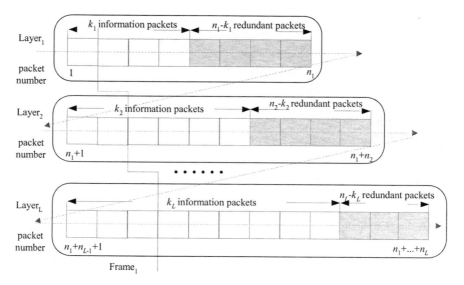

FIGURE 2.2
The packetization scheme for PFGS with UEP [43]

In the multi-layer scalable video coder such as PFGS, the impact of the residual loss probabilities of different layers on the video quality is not equal. This layered coding framework is well suited for prioritized transmission. The base layer can be assigned to a high-priority class while enhancement layers can be assigned to lower-priority classes. Since the current Internet only provides best-effort service, prioritized transmission can be achieved by applying unequal loss protection schemes to different layers. In our work, unequal loss protection is achieved by protecting different layers with different FEC codes. More specifically, strong channel-coding protection is applied to the base-layer data stream to produce a higher-priority data class while weaker channel-coding protection is applied to the subsequent enhancement layers to produce low-priority classes. The result will be that the base-layer data stream will experience a lower packet-loss probability while delivered over Internet. The packetization of PFGS with UEP is depicted in Figure 2.2. The transmission order for the packets is marked as a dashed line.

It is well known that efficient FEC codes are desirable to enable error recovery with as little overhead as possible. For the RS code that we used in our work, maximizing the FEC code rate k/n for specific network conditions is quite important to improve the protection efficiency. We should point out that, since the data packet sizes are not fixed, for a block of k data packets, the resulting $n - k$ FEC packets are all of the maximal size (denoted as *PacketLen*1). In the meanwhile, stuffing is needed for the k data packets. This will decrease the utilization efficiency of the available bandwidth.

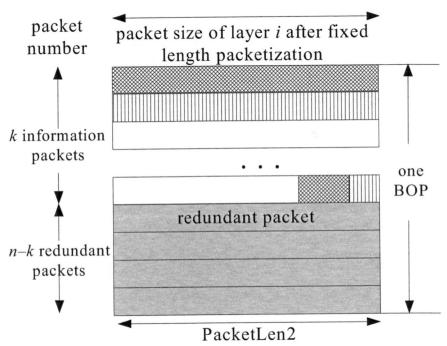

FIGURE 2.3
Generation of FEC packets using fixed-length packetization scheme [43]

To increase the bandwidth utilization, the Error Resilient Entropy Code (EREC) approach [28] is applied here (see Figure 2.3) to re-assemble different packets of data into k packets to form a Block of Packets (BOP). The basic idea of EREC is to re-organize the variable-length blocks into the EREC frame structure such that each block (slice in PFGS case) starts at a known position within the code. In this way, the decoder can independently find the start of each block. The EREC frame structure is composed of k slots of length d_i (equals to *PacketLen*2 in this case) bytes to yield total length of $T = \sum_{i=1}^{k} d_i$ bytes to transmit. The k slots of data can be transmitted consecutively without risk of any loss of synchronization. Each EREC frame can be used to transmit up to k variable-length blocks of data, provided that the total data to be coded does not exceed the total available T bits. The EREC places the variable-length blocks of data into the EREC code structure using a bit-reorganization algorithm that relies only on the ability to determine the end of each variable-length block. The details of the bit-reorganization algorithm can be found in [28].

By using the EREC approach, fixed length packetization is achieved. Small stuffing is needed in this fixed-length packetization scheme. The packet size changes from $PacketLen1$ to $PacketLen2$. The bandwidth utilization is improved approximately by $\frac{PacketLen2 - PacketLen1}{PacketLen2} \times 100\%$.

Since some error resilience mechanisms have been used in PFGS, the distortion for packet loss may just affect the slice. On the encoder side, distortion for each slice can be measured independently in advance. Let $D_S(R_S)$ stand for the source perceptual distortion-rate function. The problem becomes to find the optimal FEC scheme (k_i, n_i) for different layers to minimize the end-to-end distortion D:

$$
\begin{aligned}
\text{Minimize} \quad D = & D_S(R_S) \times P(0, R_S/S_p) + \\
& \sum_{i=1}^{m} (w_i \times \sum_{j=k_i}^{n_i} (D(i,j) \\
& \times (\sum_{l=n_i-k_i+1}^{n_i} (P(l, n_i)| \prod_{x=1}^{i-1} \sum_{y=0}^{k_x} P(y, n_x)))))),
\end{aligned}
\tag{2.6}
$$

where $D(i,j)$ represents the distortion that j^{th} packet at i^{th} layer is lost, w_i is the distortion weight for the ith layer, and m is the number of layers needed to be transmitted. Based on the decoder performance of PFGS, if the corresponding packet at any lower layers is lost, the packet of this layer is treated as lost no matter whether it is received or not.

2.2 QoS-Adaptive Resource Allocation for Scalable Video Transmission over Cellular Networks

Video is foreseen to become a key application in the Internet and mobile networks. The Third Generation (3G) wireless network is enabling video application on mobile devices and has made it feasible for visual communications over the wireless networks [6, 9], providing up to 384 kbps outdoor and 2 Mbps indoor bandwidth.

Error rate is usually very high in wireless channels, which is caused by multi-path fading, inter-symbol interference, and noise disturbances. The channel error rate varies with the changing external environment, resulting in a devastating effect on multimedia transmission. To cope with errors in wireless channel, accurate network-condition estimation [22] and effective error control [14, 38] are essential for robust video transmission. It is known that video transmission is delay-sensitive but may be tolerant to some kinds of errors. Moreover, different portions of video bit stream have different importance to the reconstructed video quality, thereby giving rise to different QoS requirements. For instance, it is intuitive that lower layers of a layered scalable video codec have higher QoS requirements than those of higher layers. Therefore, adopting different error control schemes for each portion is more appropriate for such a compressed bitstream [14]. However, channel coding introduced by error control would generate additional bit overhead and increase computational complexity. Considering the limited bandwidth in wireless networks and battery life in the mobile devices, those available resources, such as bandwidth and power, should be allocated appropriately for source and channel coding.

Most of the existing works focus on robust video transmission over 3G wireless channels [14, 40, 42]. There is, however, a scarcity of work performed on delivering multimedia over a 3G wireless network or system [1, 5]. A key issue in a 3G system or network with several layers is that it is mostly not obvious how to achieve end-to-end optimality for multimedia delivery, although a single-layer performance can reach optimum. Aiming to solve such a problem, we present, at the application layer, how to perform error control and resource allocation by taking channel/network condition into account. Specifically, we first study measurements of error rate and throughput of the 3G network. Then, resources are allocated between the source and channel coding according to the channel/network condition.

As is known, there are two basic error correction mechanisms, namely Automatic Retransmission Request (ARQ) and FEC, of which FEC has been commonly suggested for real-time applications due to the strict delay requirements. However, FEC incurs constant transmission overhead even when the channel is loss free. ARQ has been shown to be more effective than FEC [18]. But retransmission of corrupted data frames introduces additional delay, which is critical for real-time services. Hybrid FEC and ARQ schemes [41] can achieve both delay bound and rate effectiveness by limiting the number of retransmissions. However, it assumed that the maximum number of retransmissions is fixed and known *a priori*, which may not reflect the time-varying nature of delay. Meanwhile, no resource allocation between FEC and ARQ had been considered in [41]. Hybrid FEC and delay-constrained ARQ was discussed in [40, 26]; however, channel dynamics and media characteristics had not been addressed in those schemes.

Joint work on scalable video coding with UEP for wireless communication has been studied in [14]. Unequal error protection schemes improve video quality by partitioning the bit stream into different priority classes. However, a fixed error correction rate for a different priority class was adopted in [14]. This is inefficient since the characteristic of the time-varying channel has not been considered. Using scalable video compression scheme [21], it is possible to generate a single compressed bit stream such that different subsets of the stream correspond to the compressed version of the same video sequence at various rates. This is very beneficial for video delivery over time-varying networks. In [8], Cheung et al. proposed an optimal bit-allocation scheme for joint source channel coding of scalable video. In this scheme, Rate Compatible Punctured Convolutional (RCPC) and UEP are combined, but no ARQ is addressed. Meanwhile, a memoryless channel with known average channel state is assumed therein [8], which may not properly reflect the time-varying fading channel. In all the above schemes, no power constraint had been considered.

There are several related works on the power consumption in literature. Hayinga et al. [13] proposed an energy-efficient error-control scheme without considering the source side. Pan et al. [25] discussed the complexity-scalable transform coding on the source side only. Lan and Tewfik [20] considered the problem of minimizing the total consumed energy of a wireless system subject to quality-of-service constraint. Appadwedula et al. [2] proposed an efficient wireless image transmission scheme under a total power constraint. However, in [20, 2] only image is considered, and no specific error control scheme was discussed. Meanwhile, time-varying channel conditions were not taken into account in [20, 2].

To address the aforementioned issues, we present distortion-minimized resource allocation and power-minimized resource allocation with hybrid delay-constrained ARQ and UEP for video transmission over 3G network, based on the measurements of throughput and error rate for 3G wireless network. Note that our schemes are capable of dynamically adapting to the varying channel/network condition. The end-to-end architecture is illustrated in Figure 2.4. The two key components introduced in this architecture are *network throughput/error rate measurement* and *distortion/power optimized resource allocation.*

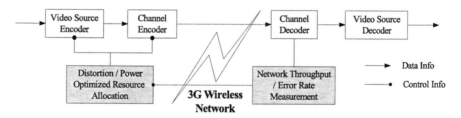

FIGURE 2.4
End-to-end architecture for video over 3G network [44]

Next, we present distortion-minimized resource allocation and power-minimized resource allocation with hybrid delay-constrained ARQ and UEP for video transmission over a 3G network.

Channel performance measurement has been discussed in [44]. Now the problem is how to efficiently utilize the limited channel capacity. According to the analysis in the previous section, both source coding and channel coding will occupy a certain portion of the resources (e.g., bits and processing power), thereby making different contributions to the end-to-end QoS, such as distortion, delay, and power consumption. The study of resource allocation in this section is to address the problem of finding the optimal distribution of resources among a set of competing subscribers (e.g., source coder and channel coder) that minimize the objective function, such as distortion or power consumption, subject to total resource constraints and/or QoS requirements.

Like many existing resource allocation algorithms [21, 27], we tackle a common case where the objective function O is the sum of an individual subscriber's objective function o_i, subject to the sum of the an individual subscriber's required resource r_i, which would not exceed the resource limit R, and/or the sum of the individual subscriber's QoS requirement q_i, which would not exceed the total QoS requirement Q. Mathematically, the problem is formulated as

$$\underset{\{q_i \ and/or \ r_i\}}{\text{Minimize}} \quad O = \sum_{i=1}^{N} o_i(q_i, r_i) \quad \text{subject to} \sum_{i=1}^{N} q_i \leq Q \ \text{and/or} \ \sum_{i=1}^{N} r_i \leq R,$$

$$(2.7)$$

where N is the number of subscribers.

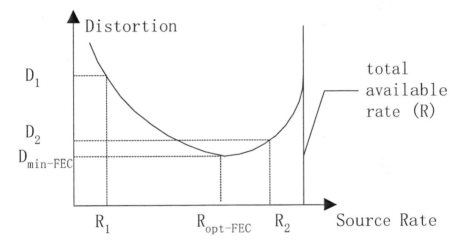

FIGURE 2.5
Rate-distortion relation with FEC scheme [44]

In this formulation, resource distribution between the PFGS source coder and the channel coder is based on the above formulation. From the rate-distortion relation analyzed in [44], it is essential to adopt some error protection schemes so as to reduce the distortion caused by channel transmission. As is known, FEC is suited for real-time communications. But varying channel condition limits its effective use, because a worst-case design may lead to a large amount of overhead. Once the channel condition is known, adaptive FEC can be adopted to meet the channel condition. Specifically, if the network condition is good, the error correction rate will be reduced. On the other hand, if the network condition is bad, the error correction rate will be

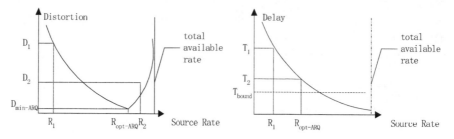

FIGURE 2.6

Rate-distortion relation with ARQ scheme and delay caused by ARQ scheme [44]

increased. As shown in Figure 2.5, there exists an optimal rate $(R_{opt-FEC})$ for the FEC scheme to achieve the minimal distortion $(D_{min-FEC})$.

Closed-loop error control techniques, such as ARQ, have been shown to be more effective than FEC. But retransmission of corrupted data frames introduces additional delay, which is critical for real-time services. As shown in Figure 2.6, there exists an optimal rate $(R_{opt-ARQ})$ for the ARQ scheme to achieve the minimal distortion $(D_{min-ARQ})$. It can be seen that $D_{min-ARQ} \leq D_{min-FEC}$. However, in real-time applications such as conferencing and streaming, the delay constraint had to be considered. When considering media's delay constraint (T_{bound}), the optimal distortion, $D_{min-ARQ}$, cannot guarantee to be achieved. Therefore, we introduced hybrid FEC and delay-constrained ARQ as the error protection scheme for multimedia delivery [38].

2.2.1 Hybrid UEP and Delay-Constrained ARQ for Scalable Video Delivery

Figure 2.7 depicts our hybrid UEP and delay-constrained ARQ scheme for scalable video delivery. In this scheme, BL and ELs are protected differently. Because BL carries the very important information, it should be transmitted in a well-controlled way to prevent the quality of reconstructed video from degrading severely. Therefore, we add strong error protection code for BL. Note that how much protection should be added to BL is based on the channel condition and available resource. As analyzed above, FEC usually incurs overhead, and the ARQ scheme is usually more efficient than FEC provided certain delay is allowed. As a result, we adopt hybrid delay-constrained ARQ and FEC for BL error protection. It works as follows. On the sender side, based on the delay constraint $D_{constrained}$ that is limited by video frame rate, current round-trip transmission time RTT, and the estimated time consumed by processing procedure $D_{processing}$, the maximum number of transmissions for current packet N_{max} can be calculated as follows:

$$N_{max} = \frac{D_{constrained} - D_{processing}}{RTT}. \tag{2.8}$$

Then, the sender determines the level of protection for each transmission such that the required residual error rate is within the desired range and the overhead is minimized.

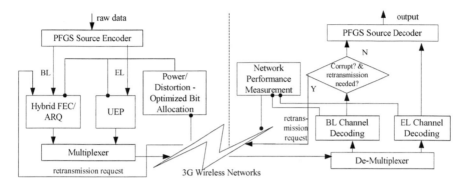

FIGURE 2.7
Block diagram of hybrid UEP/delay-constrained ARQ scheme for PFGS delivery over 3G [44]

As for ELs, different levels of error protections are added to the different layers. This is because errors in the lower layer may heavily corrupt the corresponding higher layers in the same frame, thus affecting several subsequent frames. In other words, a bit error would result in error propagation. As a result, the bandwidth for higher layers is wasted, and in the meantime, the video quality is deteriorated. Note that, in order to efficiently add error protection to ELs, the sender determines the degree of protection for each layer adapting to the current channel condition for achieving the minimal objective function under the required QoS and resource constraints.

This scheme works as illustrated in Figure 2.7. The 3G network performance is first dynamically measured. Available throughput, bit/frame/packet error rate, and some other network related information are fed back to the sender. Given the network information, optimal resource allocation is then performed to achieve the minimal objective (e.g., distortion or power consumption). The channel decoder reconstructs packets through a channel decoding process. For ELs, the output of channel decoder is directed for source decoding; while for BL, if residual error still exists, the receiver decides whether to send a retransmission request based on the delay bound of the packet. If the delay bound has been expired, the request will not be sent. Otherwise, when receiving a retransmission request, the sender only transmits a necessary higher protection part for the corresponding packet.

In summary, this presented error protection scheme is aimed to achieve the bounded delay, adaptiveness, and efficiency. However, the hybrid UEP and delay-constrained ARQ protection scheme poses a challenging resource allocation problem, because one has to consider two issues simultaneously: the tradeoff of allocation between the source and channel codes and the tradeoff between forward error protection and retransmission.

2.2.2   Distortion-Minimized Resource Allocation

It is known that channel bandwidth capacity is highly limited in wireless networks. Meanwhile, the allocation on the source side has a tradeoff between the source coding rate and the source distortion, the FEC has a tradeoff between the error protection rate and the channel distortion, and the ARQ has a tradeoff between the retransmission times and the channel distortion. Therefore, it is very important to study how to allocate the bits among the source, the FEC, and the ARQ for a given fixed bandwidth capacity so as to achieve the minimal expected end-to-end distortion.

Suppose $R(t)$ is the available bit rate at time t, $R_S(t)$, $R_{ARQ}(t)$, and $R_{FEC}(t)$ are the bit rates used for the source, the FEC, the ARQ at time t, respectively. Then the distortion-minimized resource allocation can be formulated as

$$\underset{\{R_S, R_{ARQ}, R_{FEC}\}}{\text{Minimize}} \quad D_{end-to-end} = D_S(R_S) + D_{ARQ}(R_{ARQ}) + D_{FEC}(R_{FEC})$$
$$\text{subject to} \quad R_S(t) + R_{ARQ}(t) + R_{FEC}(t) \le R(t),$$
$$(2.9)$$

where $D_S(R_S)$ is the source distortion caused by source coding rate R_S, $D_{ARQ}(R_{ARQ})$ and $D_{FEC}(R_{FEC})$ are the residual channel distortions caused by applying retransmission rate R_{ARQ} and error protection rate R_{FEC}, respectively.

The bit rate of the source side is composed of bit rate in both BL and ELs. Mathematically, it is given by

$$R_S = R_{S\_base} + \sum_{i=1}^{L} R_{S\_enh}(i), \qquad (2.10)$$

where L is the number of layers in ELs, R_{S_base} and R_{S_enh} represent the source rates of BL and of ELs, respectively.

Source distortion is composed of distortion in both BL and ELs, which can be described as

$$D_S(R_S) = D_S(R_{S\_base}) + \sum_{i=1}^{L} D_S(R_{S\_enh}(i)). \qquad (2.11)$$

Next, we will discuss the channel distortion. As discussed above, we adopt hybrid delay-constrained ARQ and FEC for BL to reduce the residual error.

It works as follows. The sender determines the degree of protection for each transmission such that the expected end-to-end distortion is minimized while satisfying the QoS requirement. Upon receiving the retransmission request for the corrupted packet, the source side will only transmit the necessary part of higher protection for the packet. Because only the protection code needs to be transmitted over the channel for re-transmission, the transmission overhead can be reduced. In this work, we use $RS(n, k)$ code for forward error correction, as mentioned before. Suppose n is fixed and let $t_i = \lfloor \frac{n-k_i}{2} \rfloor$ represent the protection level for the ith transmission. Then, the protection rate needed for the BL delivery is calculated as follows:

$$R_{ARQ} = \sum_{i=1}^{bn} R_{prot}(t_1, R_{S\_base}(i))$$

$$+ \sum_{j=2}^{N_{max}-1} \{ \sum_{i=1}^{bn} [P_{fail}(i, j-1) R_{prot}(t_j, R_{S\_base}(i))] \}, \quad (2.12)$$

$$R_{prot}(t, R_{SS}) = 2t/n \times R_{SS}, \quad (2.13)$$

$$P_{fail} = \prod_{l=1}^{j} P_{fail,packet}(i, l), \quad (2.14)$$

and

$$P_{fail,packet}(i, j) = 1 - \sum_{x=0}^{t_j} \{ \sum_{y=0}^{x} [\binom{n}{y} P_S(i)^y (1 - P_S(i))^{n-y}] \}, \quad (2.15)$$

where b_n is the number of source packets needed to be transmitted, $R_{prot}(t, R_{SS})$ is the bit rate needed for protecting R_{SS} at level t, $P_{fail}(i, j)$ is the probability of the ith packet failed in the past j times retransmission, $P_{fail,packet}(i, j)$ is the probability of the ith packet that is failed in the jth retransmission, and $p_S(i)$ is the probability of symbol failure of the ith packet.

After hybrid FEC and delay-constrained ARQ protection for BL, only those blocks that cannot be recovered will cause the additional channel distortion. Thus, the channel distortion of BL can be described as

$$D(R_{ARQ}) = \sum_{i=0}^{bn} [P_{fail}(i, N_{max} - 1) \times D_c(i)], \quad (2.16)$$

where $D_c(i)$ is the channel distortion caused by the loss of packet i.

Now we analyze the channel distortion in ELs. Considering the dependency among layers, UEP is applied for the ELs. Similar to the BL, we use $t_i = \lfloor \frac{n-k_i}{2} \rfloor$ to represent the protection level for the ith layer. The protection rate needed for the ELs delivery is then represented as follows:

$$R_{FEC} = \sum_{i=1}^{L} R_{prot}(t_i, R_{S\_enh}(i)), \tag{2.17}$$

where L is the number of layers needed to be transmitted, and $R_{prot}(t, R_{ss})$ is the bit rate needed for protecting R_{ss} at level t, which had been defined in (2.13). Then, the channel distortion of ELs after UEP can be expressed as

$$D(R_{FEC}) = \sum_{i=1}^{L} [\sum_{j=1}^{bn_i} (P_{fail,layer}(i,j) \times D_c(j))], \tag{2.18}$$

$$P_{fail,layer}(i,j) = P_{fail,packet,layer}(i,j)[\prod_{m=1}^{i-1} (1 - P_{fail,packet,layer}(m,j))], \tag{2.19}$$

and

$$P_{fail,packet,layer}(i,j) = 1 - \sum_{x=0}^{t_i} \{\sum_{y=0}^{x} [\binom{n}{y} P_S(j)^y (1 - P_S(j))^{n-y}]\}, \tag{2.20}$$

where bn_i is the number of source packets needed to be transmitted in the ith layer, $P_{fail,layer}(i,j)$ is the probability of which the jth packet in the ith layer is corrupted while the corresponding packets in the previous layers are correct, and $P_{fail,packet,layer}(i,j)$ is the probability of that the jth packet is corrupted in the ith layer.

Substituting Eqns. (2.12, 2.16, 2.17, 2.18) into Eqn. (2.9), the distortion-minimized resource allocation for scalable video delivery can be solved given the total bit budget $R(t)$.

Figure 2.8 depicts the corresponding rate-distortion relation of our proposed hybrid UEP and delay-constrained ARQ scheme. Based on the above analysis and from Figure 2.8, we can see $D_{min-ARQ} \leq D_{min} \leq D_{min-FEC}$. In the meantime, delay bound of media (T_{bound}) is satisfied.

2.2.3   Power-Minimized Resource Allocation

Besides the channel bandwidth capacity, another highly limited resource in wireless networks is power, which includes the transmitter power and receiver power. In this formulation, we only consider the receiver power in the mobile devices, which consists of receiving power, source decoding power and channel decoding power. It is observed that both the source and the channel have a tradeoff between the coding rate and the processing power consumption. Thus, the power-optimized resource allocation problem can be formulated as: given the fixed bandwidth capacity, how we allocate bits among the source, the FEC, and the ARQ so as to achieve the minimum power consumption under

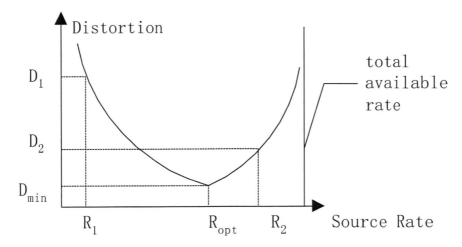

FIGURE 2.8
Rate-distortion relation with hybrid UEP/delay-constrained ARQ scheme [44]

the desired end-to-end distortion range. Let $R(t)$ represent the available bit rate at time t, $R_S(t)$, $R_{ARQ}(t)$, and $R_{FEC}(t)$ represent the bit rate used for the source, the FEC, the ARQ at time t, respectively, and $D(t)$ represent the tolerable distortion at time t. Then, the power-minimized resource allocation can be described as

$$
\begin{aligned}
\underset{\{R_S, R_{ARQ}, R_{FEC}\}}{\text{Minimize}} \quad PC = \; & PC_{rec,S}(R_S) + PC_{rec,ARQ}(R_{ARQ}) \\
& + PC_{rec,FEC}(R_{FEC}) + PC_S(R_S) \\
& + PC_{ARQ}(R_{ARQ}) + PC_{FEC}(R_{FEC}) \\
\text{subject to} \quad & D_S(R_S) + D_{ARQ}(R_{ARQ}) + D_{FEC}(R_{FEC}) \le D(t) \\
\text{and} \quad & R_S(t) + R_{ARQ}(t) + R_{FEC}(t) \le R(t),
\end{aligned}
\tag{2.21}
$$

where $PC_{rec,S}(R_S)$, $PC_{rec,ARQ}(R_{ARQ})$, $PC_{rec,FEC}(R_{FEC})$ are the power consumed for receiving the source, the ARQ, and the FEC, respectively, and $PC_S(R_S)$, $PC_{ARQ}(R_{ARQ})$, $PC_{FEC}(R_{FEC})$ are consumed power for the source coding, the ARQ, and the FEC, respectively.

As analyzed in [44], source decoding and channel decoding had different power consumptions. For the source part, the receiving power is composed of receiving powers for both BL and ELs. Mathematically,

$$
PC_{rec,S}(R_S) = \sum_{i=1}^{bn} [\rho_{rec} \times (R_{S\_base}(i))] + \sum_{j=1}^{L} \{ \sum_{i=1}^{bn_j} [\rho_{rec} \times (R_{S\_enh}(i,j))] \},
\tag{2.22}
$$

where bn_j is the number of blocks in the jth layer, and ρ_{rec} is the power consumed for per bit transmission. The consumed processing power for the source part is related to the source decoding rate, which is denoted as

$$PC_S(R_S) = \rho s(R_S) = \sum_{i=1}^{bn} \rho s(R_{S\_base}(i)) + \sum_{j=1}^{L} \{\sum_{i=1}^{bn_j} \rho s(R_{S\_enh}(i,j))\} \quad (2.23)$$

where $\rho s(.)$ can be obtained from Figure 8 in [44].

As for the channel part, the consumed processing power is related to both the source decoding rate and the channel protection rate, which is represented as $PC_{FEC}(R_s, R_{FEC}) = \rho s(R_s, R_{FEC}) = \rho c(R_s, t)$, where t is the error protection level, and $\rho c(.)$ can be obtained from Figure 9 in [44].

In our hybrid delay-constrained ARQ and FEC scheme that used for BL, any corrupted packet is allowed to be transmitted at most N_{max} times. Once receiving the retransmission request, a higher protection level is determined by the sender to achieve the desired video quality. On the sender side, only higher protection code will be transmitted to the receiver. While on the receiver side, different channel decoding would be performed after each transmission. Thus, the receiving power consumption for the BL is formulated as

$$\begin{aligned} PC_{rec,ARQ}(R_{ARQ}) &= \sum_{i=1}^{bn} [\rho_{rec} \times R_{prot}(t_1, R_{s\_base}(i))] \\ &+ \sum_{j=2}^{N_{max}-1} \{\sum_{i=1}^{bn} [P_{fail}(i,j-1) \times \rho_{rec} \times R_{prot}(t_j, R_{s\_base}(i))]\}, \end{aligned} \quad (2.24)$$

where t_i is the error protection level for the ith retransmission. Similarly, the processing power consumption for the BL is represented as

$$\begin{aligned} PC_{ARQ}(R_{ARQ}) &= \sum_{i=1}^{bn} \rho c(R_{s\_base}(i), t_1) \\ &+ \sum_{j=2}^{N_{max}-1} \{\sum_{i=1}^{bn} [P_{fail}(i,j-1) \times \rho c(R_{s\_base}(i), t_j)]\}. \end{aligned} \quad (2.25)$$

As discussed before, UEP is applied to ELs. To be specific, different channel protection bits will be transmitted for different layers on the sender side; while different channel decoding will be performed for different layers on the receiver side. The receiving power consumption for ELs is represented as

$$PC_{rec,FEC}(R_{FEC}) = \sum_{j=1}^{L} \{\sum_{i=1}^{bn_j} [\rho_{rec} \times R_{prot}(t_j, R_{s\_enh}(i,j))]\}. \quad (2.26)$$

Similarly, the processing power consumption for ELs is expressed as

$$PC_{FEC}(R_{FEC}) = \sum_{j=1}^{L} \{\sum_{i=1}^{bn_j} [\rho c(R_{s\_enh}(i,j), t_j)]\}. \quad (2.27)$$

Substituting Eqns. (2.22–2.27) into Eqn. (2.21), the power-minimized resource allocation for scalable video delivery can be solved, given the total bit budget $R(t)$ and the desired distortion range $D(t)$. Note that optimization methods, such as Lagrange multiplier and penalty function methods, can be used to solve the constrained non-linear optimization problem.

2.3 Power-Minimized Joint Power Control and Resource Allocation for Video Communications over Wireless Channels

The traditional relationship between rate and distortion assumes that power is infinite. For mobile multimedia communication using handheld devices, power consumption usually is finite due to the limited battery life of mobile devices. Motivated by this, we explore the relationship between power and rate given distortion in wireless communications including sender and the receiver (note that power-minimized resource allocation in the above section only considers the receiver power consumption). Specifically, we want to optimally determine the power in source coding, channel coding, and transmission, respectively, and rate allocated for both source and channel, for a given distortion.

Video over wireless networks has undergone enormous development, due to continuing growth of wireless communication. However, wireless video communications face several challenges including bandwidth requirement and battery lifetime constraints. Minimizing average power consumption and keeping the QoS at Mobile Station (MS) usually conflict each other. Achieving the best multimedia quality at MS usually consumes more power. Moreover, multipath fading in radio channels necessitates the use of large transmit power and complex signal processing algorithms, which largely reduces the battery life. In addition, to combat the Multiple Access Interference (MAI) from other users in the same cell, the transmission power is further increased.

Power control and Joint Source-Channel Coding (JSCC) are two effective approaches to supporting QoS for robust video communications over wireless networks. Power control is performed from a group point of view by controlling the transmission power and spreading gain (transmission rate) of a group of users, while JSCC is conducted to effectively combat the errors that occurred during the transmission.

From an individual point of view, the approach to allocate different bit rates for source and channel units takes effect on minimizing the total power consumption under a fixed bandwidth constraint. In today's cellular phone, the bulk of the power is consumed in the power amplifier that generates the transmit power. However, as we move toward an era of micro-cells and pico-cells, power consumption of the processing power for multimedia data becomes comparable to that of the power amplifier. Such an approach can be viewed as an extension of the joint source-channel bit-allocation scheme presented in [8], which considers the efficient joint design of the source and channel coder to minimize the overall distortion under a fixed bandwidth constraint. There are some existing works related to minimizing the power consumption while performing JSCC for a single user. To be specific, a low-power communication system for image transmission was investigated [11]. Ji proposed a power-optimized JSCC approach for video communications over wireless channels

[17]. However, in [11, 17] the interference to other users when performing the optimization of power consumption was not considered.

From a group point of view, power control adjusts a group of users' transmission power to maintain their QoS requirements and solve optimization problems, such as transmitter power minimization, network capacity maximization, and optimal resource allocation. Recently, the focus has been on adjusting transmitter powers to maintain a required Signal-to-Interference Ratio (SIR) threshold for each network link using the least possible power. It can also be referred to as an approach to resource management based on the power control technique discussed in [31, 35, 19], where it is formulated as a constrained optimization problem to minimize the total transmission power or maximize the total rate constrained by the SIR and bandwidth requirements. Nevertheless, processing power and JSCC have not been taken into account in [31, 35, 19].

To jointly consider power control and JSCC, we present an approach that minimizes the processing-and-transmission power consumption for a single user and all users in a single-cell Code Division Multiple Access (CDMA) system, respectively. Therein, joint consideration of power control and source-channel coding is performed to achieve the minimal total power consumption, meanwhile, maintaining the desired QoS for video transmission. Furthermore, we develop an iterative solution to MS-BS interactive framework to solve the optimization based on Wideband Code Division Multiple Access (W-CDMA) infrastructure.

An architecture for power-minimized video communications over wireless channel is shown in Figure 2.9. The key components of the architecture include a *network-aware power consumption optimizer (NAPCO), power-reconfigurable video encoder, UEP channel encoder,* and *power amplifier.*

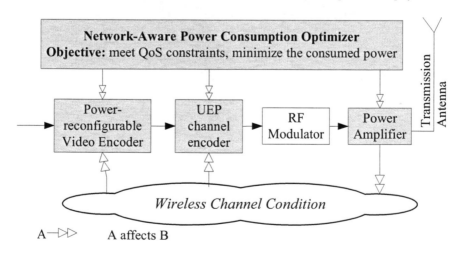

FIGURE 2.9
Architecture for joint power control and rate control in MS [45]

In order to minimize the total power consumption while satisfying the QoS requirements for each user, NAPCO is responsible for understanding the current channel status and periodically adjusting the video rate, protection rate, transmission power update step, and target E_b/N_0. Our joint power control and rate control is to allocate the bits between source and channel by minimizing the total power consumptions of the source, channel and transmission. First, NAPCO analyzes power consumption on the source side. In the video encoder, Motion Estimation (ME) is most computationally intensive (up to about 50% of the entire system). The computation complexity and the residual distortion of ME vary in a large range with different block matching precisions. In this approach, a partial-distortion-measure based hierarchical block motion estimation algorithm [7] is adopted in a power-reconfigurable video encoder to provide several power consumption levels with different Mean Square Error (MSE). Second, NAPCO monitors the power consumption spent on the channel coding. In this approach, a UEP scheme based on RS codes is used to protect the compressed video bitstream. Different portions of compressed bitstreams have different importance to the quality of the reconstructed video. The computation complexity and video transmission distortion vary with different levels of error protection codes. Third, NAPCO analyzes power consumption of transmission. The transmission power of each user is related to the transmission bit energy and the total bit rate. Adjusting the transmission bit energy of an MS will not only affect its own channel condition, but will also change the channel status of other users in the same cell.

From the above analysis, we can see that all those three components have relationships with its correspondent bit rate and channel condition. Once the total power consumption consumed in source and channel coding, as well as transmission, NAPCO allocates bits between source coding and channel coding according to the channel condition and QoS requirement.

Next, we present power-minimized rate allocation for a single user and a group of users, respectively.

2.3.1   Power-Rate Relationship and Power-Minimized Rate Allocation for a Single User

In the video transmission scheme, the total power consumed by a given user consists of transmission power and processing power. Processing power is mainly determined by computational costs for source coding and channel coding. Transmission power, on the other hand, depends on the bit energy and total bit rate to deliver. Both transmission power and processing power should be controlled so as to adapt to the changing channel conditions. In this sub-section, we assume that the multiple access interferences are successfully removed from the expected receiving signal, which can be approximated by utilizing synchronized orthogonal spreading code or optimal multiuser detection. In this way, the minimal total power consumption of all users can be achieved by minimizing each user's power consumption, as the transmission power is not required to combat the interference of other users. Hence, we first investigate the power-consumption minimization for a single user in this sub-section.

Having analyzed the power consumed in each individual component above, now we allocate the available rate between source and channel, as well as adjusting the source power levels and the transmission power, so as to minimize the total power consumption while satisfying the QoS requirements, such as the user's uplink distortion and total bit rate constraint. Specifically, the power-optimized rate allocation problem can be formulated as

$$\begin{array}{ll} \underset{\{R_s, R_c, \gamma, \Phi\}}{\text{Minimize}} & P_s(\Phi, R_s) + P_c(R_c) + \varepsilon_t(P_t(\gamma, R_s + R_c)) \\ \text{subject to} & D_u \leq D_0 \quad \text{and} \quad R_s + R_c \leq R_0. \end{array} \tag{2.28}$$

where D_u and D_0 respectively represent the expected uplink distortion and maximal tolerable distortion, which can be derived from the QoS requirements. R_s, R_c and R_0 are the source rate and channel coding rate and the constraint of the total rate, respectively. P_s, P_c, and P_t are the power consumption for video source encoding, channel UEP, and data transmission, respectively. Let γ be the target E_b/N_0's. The transmission power can be calculated given γ and the path loss from the mobile station to the base station.

For the given user, the expected uplink distortion, D_u, is composed of the source distortion (D_s) and the channel distortion (D_c). The source distortion is caused by the searching range in the ME algorithm and video rate control; while the channel distortion results from the channel transmission error. Mathematically,

$$D_u = D_s(R_s) + \sum_{m=1}^{N_{msp}} P_{fail}(m) \times D_c(m) + \sum_{n=1}^{N_{lsp}} P_{fail}(n) \times D_c(n), \tag{2.29}$$

where N_{msp} is the number of blocks in MSP, N_{lsp} is the number of blocks in LSP, $P_{fail}(m)$ is the failure probability of the mth block, and $D_c(m)$ is the channel distortion caused by the failure of the mth block.

Considering a fixed target E_b/N_0 and source power level for a given user, the optimization problem (2.28) becomes to search a minimal total rate while satisfying the distortion requirement under the rate constraint, as illustrated in Figure 2.10. Conventional JSCC can be used here to find the optimal set of (R_s, R_c) to obtain the minimal total rate. Taking the various target E_b/N_0 into consideration, Figure 2.11 shows the total power consumption of a given user as a function of source rate, total rate, and E_b/N_0 requirement. When the target E_b/N_0 is low, more protection bits are necessary to combat the high BER of the channel. On the contrary, it needs fewer bits to perform channel protection as the target E_b/N_0 is high, which results in a low BER of the channel. After all, for the above two cases, the product of total rate and target E_b/N_0 remains a large value. It indicates a curve with convex characteristics as Figure 2.11 illustrates. The optimal set of (R_s, R_c, γ) can be obtained from the surface of the curve.

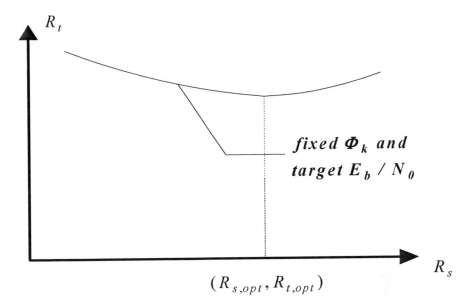

FIGURE 2.10
The total rate of a given user when fixing Φ_k and target E_b/N_0 [45]

Moreover, considering the fixed target E_b/N_0 but taking different source power levels into account, Figure 2.12 shows the curve of the total rates with different Φ. Therefore, the optimal set (R_s, R_c) of minimizing the mobile's power consumption is chosen from the sub-optimal sets at different Φ. As the various target E_b/N_0 is considered, the optimal set of (R_s, R_c, γ, Φ) can be achieved at the surfaces of several 3-dimentional curves according to varying Φ^i, which are similar to the curve shown in Figure 2.10.

In order to obtain the optimal sets of the optimization problem, channel-adaptive power and rate control are adopted in our proposed system to minimize the total power consumption, which outperforms the fixed power-controlled system.

2.3.2 Power-Rate Relationship and Power-Minimization Rate Allocation for a Group of Users

As the existing W-CDMA standards illustrate, no specific multiuser detection algorithm is included and the synchronization of uplink transmission is difficult to implement in practical W-CDMA systems. However, power control is considered as a primary approach of transmitter power minimization, network capacity maximization, and optimal resource allocation. In this subsection, power control is further studied to minimize the total power consumption of the processing unit and power amplifier of a group of users. Due to the mul-

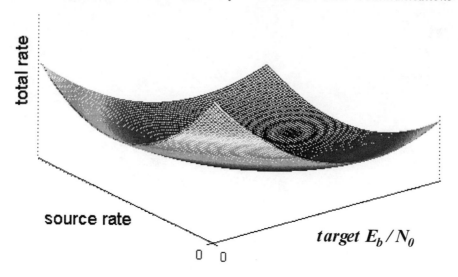

FIGURE 2.11
The total power consumption of given user as a function of source rate, total rate, and E_b/N_0 requirement [45]

tiple access interference, the global minimization of power consumption must be investigated from a group point of view as the transmission power of the group of users are correlated to each other. Moreover, the capacity constraint of the uplink system should be taken into account.

On the basis of the power minimization for a single user analyzed in Section 2.3.1, we extend the optimization problem to the N-dimension, where N is the number of users in the cell. We now adjust the power control step size, target SIR, source rate, protection levels and source power levels to minimize the total power consumption while satisfying the QoS requirements for each mobile. The total power consumption is composed of source coding power, channel coding power and transmission power. Hence, considering a single cell system, the power-optimized rate allocation problem for all mobiles can be formulated as

$$\underset{\{R_s^i, R_c^i, \Gamma, \Phi^i\}}{\text{Minimize}} \quad \sum_{i=1}^{N} \left(P_s^i(\Phi^i, R_s^i) + P_c^i(R_c^i) + \varepsilon_t^i(P_t^i(\Gamma, \mathbf{R})) \right) \tag{2.30}$$
$$\text{subject to} \quad D_u^i \leq D_0^i, \quad i = 1, 2, \dots, N,$$

where the superscript i represents the ith user, D_u^i and D_0^i respectively represent the expected uplink distortion and maximal tolerable distortion, which can be derived from the QoS requirements. R_s^i and R_c^i are the source rate and channel coding rate, respectively. P_s^i, P_c^i, and P_t^i are the power consumption for video source encoding, channel UEP, and data transmission, respectively. Define the vector of the target E_b/N_0's $\Gamma = [\gamma^1, \gamma^2, \dots, \gamma^N]$ and the vector of transmitted rates to be $\mathbf{R} = [R_t^1, R_t^2, \dots, R_t^N]$.

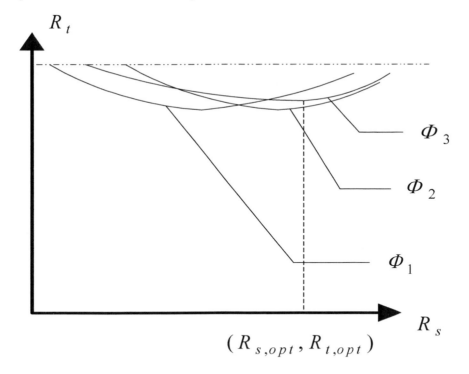

FIGURE 2.12
The total rate of given user with various Φ_k and fixed target E_b/N_0 [45]

Let $r'_i = R^i_t \gamma^i_t$, we can obtain P^i_t from the following equation [31]

$$(\frac{W}{r'_i} + 1)h_i P^i_t [1 - \sum_{j=1}^{N} \frac{1}{(\frac{W}{r'_j} + 1)}] = \eta_0 W \quad i = 1, 2, \ldots, N, \quad (2.31)$$

where W and h_i are respectively the total bandwidth for all users and the channel gain of the ith user. It can be seen from (2.29) that it is necessary to be aware of $r'_i s$ of other users when calculating the transmission power of the current user. Moreover, if every user's r'_i is minimized, the transmission power of each user reaches the minimum simultaneously. We can define

$$\Psi = [1 - \sum_{j=1}^{N} \frac{1}{(\frac{W}{r'_j} + 1)}], \quad (2.32)$$

which is calculated in the BS and inform each MS through a pilot channel of the interference of other users. Considering the fixed r'_i, that is to fix Ψ, each user can minimize its power consumption according to the analysis of *Part A* in this section. Moreover, varying from a N-dimension space, from where we can get the optimal set of r'_i. If the r'_i of some users is increased, accordingly decreasing the Ψ, other users' transmission power is enhanced. At the processing unit side, we could increase the source power level to diminish source rate and add more protection bits to decrease the value of target E_b/N_0, which results in reducing the transmission power. Meanwhile, the processing power consumption is enlarged. Then, the minimized total power consumption for each user can be achieved by trading off between processing power and transmission power. On the contrary, decreasing the r_i' of some users, accordingly increasing the Ψ, other users' transmission power is reduced. At the processing unit side, we could reduce the source power level and add fewer protection bits, which results in increased transmission power. At the same time, the processing power consumption is diminished. The minimal power consumption may be achieved in this way.

Furthermore, notice that influences from different users on the capacity of the system are different. Considering the capacity constraint of a power-control system with transmission power constraint, p_i, the following inequality is obtained [31]:

$$\sum_{j=1}^{N} \frac{1}{(\frac{W}{r_j'} + 1)} \leq 1 - \frac{\eta_0 W}{\min_i[p_i h_i(\frac{W}{r_i'} + 1)]} \quad i = 1, 2, \ldots, N. \tag{2.33}$$

This inequality manifests that users that have low power budget, who are far away, need high rates and quality of service that put a limit on capacity. Consequently, we define a weight for each user to represent its influence on the capacity in the following:

$$w_i = \frac{p_i h_i}{r_i'} \quad i = 1, 2, \ldots, N. \tag{2.34}$$

The w_i is useful to weigh the searching direction in the N-dimension space. When we minimize the total power consumption, introducing this weight results in satisfying the capacity constraint more efficiently.

As analyzed above, to solve the global power and rate allocation problem (2.30), one approach is to apply a centralized algorithm which considers all users' statistics simultaneously. The advantage of this approach is that it does not require any iterative steps for achieving the optimum; however, it needs large computational complexity and high communication requirements. On the other hand, we can utilize a distributed solution that deals with this N-dimension minimization problem by using an iterative, one-dimensional (1-D) search algorithm, in which we optimize along each user's total power consumption iteratively until reaching the minimum. This solution can decrease the complexity of solving this optimization problem and proceed in a distributed way. But it may only achieve local optimal or sub optimal sets of the N-dimension space.

2.4 Summary

In this chapter, we have examined various techniques for optimal resource allocation for scalable video communications over the Internet or wireless networks. Specifically, we have investigated the following four topics 1) network-adaptive resource allocation for scalable video streaming over the Internet, in which we have presented a novel resource allocation scheme for multiple media streams to achieve end-to-end optimal quality according to the estimated network bandwidth and media rate-distortion functions; 2) QoS-adaptive resource allocation for scalable video transmission over cellular networks, in which we have presented distortion-minimized resource allocation scheme and power-minimized resource allocation scheme with hybrid delay-constrained ARQ and UEP, for video transmission over 3G networks; and 3) power-minimized joint power control and resource allocation for video communications over wireless channels, in which we have presented an approach that minimizes the processing-and-transmission power consumption for a single user and all users in a single-cell CDMA system, respectively.

Acknowledgment

Dr. Wenwu Zhu would like to acknowledge Prof. Qian Zhang of Hong Kong University of Science and Technology for her contributions to this chapter.

Bibliography

[1] v.3.0.1 3G TS 23.101. General UMTS Architecture, 1999.

[2] S. Appadwedula, M. Goel, D.L. Jones, K. Ramchandran, and NR Shanbhag. Efficient wireless image transmission under a total power constraint. In *Multimedia Signal Processing, 1998 IEEE Second Workshop on*, pages 573–578. IEEE.

[3] R. Braden, L. Zhang, S. Berson, S. Herzog, and S. Jamin. *Resource ReSerVation Protocol:(RSVP); Version 1 Functional Specification*. Univ. of Michigan, 1997.

[4] S. Cen, C. Pu, and J. Walpole. Flow and congestion control for internet media streaming applications. In *Proceedings of Multimedia Computing and Networking*, pages 1–14. Citeseer, 1998.

[5] L.F. Chang and X.X. Qiu. Wireless internet: networking and protocol aspects. In *wireless Internet course notes.* IEEE PIMRC 2000, Sept. 2000.

[6] P. Chaudhury, W. Mohr, and S. Onoe. The 3gpp proposal for imt-2000. *Communications Magazine, IEEE*, 37(12):72–81, 1999.

[7] C.K. Cheung and L.M. Po. A hierarchical block motion estimation algorithm using partial distortion measure. In *Image Processing, 1997. Proceedings., International Conference*, volume 3, pages 606–609. IEEE, 1997.

[8] G. Cheung and A. Zakhor. Bit allocation for joint source/channel coding of scalable video. *Image Processing, IEEE Transactions on*, 9(3):340–356, 2000.

[9] E. Dahlman, P. Beming, J. Knutsson, F. Ovesjo, M. Persson, and C. Roobol. WCDMA – the radio interface for future mobile multimedia communications. *Vehicular Technology, IEEE Transactions on*, 47(4):1105–1118, 1998.

[10] M. Eckert and J. I. Ronda. Bit-rate allocation in multi-object video coding. ISO/IEC JTC1 / SC29 / WG11 MPEG98/m3757, Dublin, Ireland.

[11] M. Goel, S. Appadwedula, NR Shambhag, K. Ramchandran, and D.L. Jones. A low-power multimedia communication system for indoor wireless applications. In *Signal Processing Systems, 1999. SiPS 99. 1999 IEEE Workshop on*, pages 473–482. IEEE, 1999.

[12] MPEG Video Group. Text of iso/iec 14496-2 mpeg-4 video fgs vm 2.0. Doc. ISO/IEC JTC1/SC29/WG11 N2926, Melbourne, October 1999.

[13] P.J.M. Hayinga. Energy efficiency of error correction on wireless systems. In *Wireless Communications and Networking Conference, 1999. WCNC. 1999 IEEE*, pages 616–620. IEEE, 1999.

[14] U. Horn, B. Girod, and B. Belzer. Scalable video coding for multimedia applications and robust transmission over wireless channels. In *7th International Workshop on Packet Video*. Citeseer, 1996.

[15] U. Horn, K. Stuhlmuller, M. Link, and B. Girod. Robust internet video transmission based on scalable coding and unequal error protection. *Signal Processing: Image Communication*, 15(1-2):77–94, 1999.

[16] S. Jacobs and A. Eleftheriadis. Streaming video using dynamic rate shaping and tcp congestion control* 1. *Journal of Visual Communication and Image Representation*, 9(3):211–222, 1998.

[17] Z. Ji, Q. Zhang, W. Zhu, and Y.Q. Zhang. End-to-end power-optimized video communication over wireless channels. In *Multimedia Signal Processing, 2001 IEEE Fourth Workshop on*, pages 447–452. IEEE, 2001.

[18] M. Khansari, A. Jalah, E. Dubois, and P. Mermelstein. Robust low bit-rate video transmission over wireless access systems. In *Communications, 1994. ICC'94, SUPERCOMM/ICC'94, Conference Record, 'Serving Humanity Through Communications.' IEEE International Conference*, pages 571–575. IEEE, 1994.

[19] S.L. Kim, Z. Rosberg, and J. Zander. Combined power control and transmission rate selection in cellular networks. In *Vehicular Technology Conference, 1999. VTC 1999-Fall. IEEE VTS 50th*, volume 3, pages 1653–1657. IEEE, 1999.

[20] T.H. Lan and A.H. Tewfik. Adaptive low power multimedia wireless communications. In *Multimedia Signal Processing, 1997, IEEE First Workshop on*, pages 377–382. IEEE, 1997.

[21] S. Li, F. Wu, and Y.Q. Zhang. Study of a new approach to improve fgs video coding efficiency. *ISO/IEC JTC1/SC29/WG11, MPEG99/ M*, 1999.

[22] X. Mestre, M. Najar, and J.R. Fonollosa. Joint beamforming and channel estimation for pilot-aided wcdma systems. In *Acoustics, Speech, and Signal Processing, 2000. ICASSP'00. Proceedings. 2000 IEEE International Conference*, volume 5, pages 2577–2580. IEEE, 2000.

[23] J. Padhye, V. Firoiu, D. Towsley, and J. Kurose. Modeling TCP throughput: A simple model and its empirical validation. In *ACM SIGCOMM Computer Communication Review*, volume 28, pages 303–314. ACM, 1998.

[24] J. Padhye, J. Kurose, D. Towsley, and R. Koodli. A model based TCP-friendly rate control protocol. In *Proceedings of NOSSDAV99*. Citeseer, 1999.

[25] W. Pan and A. Ortega. Complexity-scalable transform coding using variable complexity algorithms. In *Data Compression Conference, 2000. Proceedings. DCC 2000*, pages 263–272. IEEE, 2000.

[26] R. Puri, K. Ramchandran, and A. Ortega. Joint source channel coding with hybrid ARQ/FEC for robust video transmission. In *IEEE Multimedia Signal Processing Workshop*, 1998.

[27] L. Qian, D.L. Jones, K. Ramchandran, and S. Appadwedula. A general joint source-channel matching method for wireless video transmission. In *Data Compression Conference, 1999. Proceedings. DCC'99*, pages 414–423. IEEE, 1999.

[28] D.W. Redmill and N.G. Kingsbury. The EREC: an error-resilient technique for coding variable-length blocks of data. *Image Processing, IEEE Transactions on*, 5(4):565–574, 1996.

[29] R. Rejaie, M. Handley, and D. Estrin. Quality adaptation for congestion controlled video playback over the internet. In *ACM SIGCOMM Computer Communication Review*, volume 29, pages 189–200. ACM, 1999.

[30] I. Rhee, V. Ozdemir, and Y. Yi. Tear: TCP emulation at receivers–flow control for multimedia streaming. 2000.

[31] A. Sampath, P. Sarath Kumar, and J.M. Holtzman. Power control and resource management for a multimedia CDMA wireless system. In *Personal, Indoor and Mobile Radio Communications, 1995. PIMRC'95.Wireless: Merging onto the Information Superhighway, Sixth IEEE International Symposium*, volume 1, pages 21–25. IEEE, 1995.

[32] H.R. Shao, W. Zhu, and Y.Q. Zhang. User-aware object-based video transmission over the next generation internet. *Signal Processing: Image Communication*, 16(8):763–784, 2001.

[33] J. Shin, J.W. Kim, and C.C.J. Kuo. Content-based packet video forwarding mechanism in differentiated service networks. In *Proceedings of the Packet Video Workshop*. Citeseer, 2000.

[34] D. Sisalem and H. Schulzrinne. The loss-delay based adjustment algorithm: A TCP-friendly adaptation scheme. In *Workshop on Network and Operating System Support for Digital Audio and Video*. Citeseer, 1998.

[35] M. Soleimanipour, W. Zhuang, and G.H. Freeman. Modeling and resource allocation in wireless multimedia CDMA systems. In *Vehicular Technology Conference, 1998. VTC 98. 48th IEEE*, volume 2, pages 1279–1283. IEEE, 1998.

[36] W.T. Tan and A. Zakhor. Real-time internet video using error resilient scalable compression and TCP-friendly transport protocol. *Multimedia, IEEE Transactions on*, 1(2):172–186, 1999.

[37] A. Vetro, H. Sun, and Y. Wang. Mpeg-4 rate control for multiple video objects. *Circuits and Systems for Video Technology, IEEE Transactions on*, 9(1):186–199, 1999.

[38] G. Wang, Q. Zhang, W. Zhu, and Y.Q. Zhang. Channel-adaptive error control for scalable video over wireless channel. In *IEEE MoMuc*, pages 2000–2010, 2000.

[39] L. Wang and A. Vincent. Joint coding for multi-program transmission. In *Image Processing, 1996. Proceedings.* volume 1, pages 425–428. IEEE, 1996.

[40] D. Wu, Y.T. Hou, Y.Q. Zhang, W. Zhu, and HJ Chao. Adaptive QoS control for mpeg-4 video communication over wireless channels. In *Circuits and Systems, 2000. Proceedings. ISCAS 2000 Geneva.* volume 1, pages 48–51. IEEE, 2000.

[41] Q. Zhang and S.A. Kassam. Hybrid arq with selective combining for fading channels. *Selected Areas in Communications, IEEE Journal on Selected Areas in Communications*, 17(5):867–880, 1999.

[42] Q. Zhang, W. Zhu, G.J. Wang, and Y.Q. Zhang. Resource allocation with adaptive QoS for multimedia transmission over w-cdma channels. In *Wireless Communications and Networking Conference, 2000. WCNC. 2000 IEEE*, volume 1, pages 179–184. IEEE, 2000.

[43] W. Zhang, Q. Zhu and Y.Q. Zhang. Resource allocation for multimedia streaming over the internet. *IEEE Trans. on Multimedia*, 3(3):339–355, 2001.

[44] W. Zhang, Q. Zhu and Y.Q. Zhang. Channel-adaptive resource allocation for video transmission over 3G wireless network. *IEEE Trans. on Circuit and System for Video Technology*, 14(8):1049–1063, 2004.

[45] Z. Zhu W. Zhang, Q. Ji and Y.Q. Zhang. Power-Minimized Bit Allocation for Video Communication Over Wireless Channels. *IEEE Trans. on Circuit and System for Video Technology*, 12(6):398–410, 2002.

3

Optimal Resource Allocation for P2P Streaming Systems

CONTENTS

With advances in the broadband Internet access technology and the coding techniques, video streaming services have become increasingly popular. Traditionally, video streaming services are deployed in the client/server architecture. However, this centralized architecture cannot provide streaming to a large population of users due to the limited upload capacity from the server.

Peer-to-Peer (P2P) technology has recently become a promising approach to provide live streaming services or Video-on-Demand (VoD) services to a huge number of the concurrent users over a global area. P2P streaming systems can be classified into P2P live streaming systems and P2P VoD systems. In a P2P live streaming system, a live video is disseminated to all users in real time. The video playbacks on all users are synchronized. In a P2P VoD system, users can choose any video they like and start to watch it at any time. The playbacks of the same video by different users are not synchronized.

A P2P streaming system is a distributed and collaborative system, in which peers contribute their resources to the system. The system performance is greatly dependent on the resource utilization of each peer. Therefore, we will examine the techniques for optimal resource allocation for P2P streaming systems in this chapter. Specifically, we will present the the following two topics: 1) distributed throughput maximization for scalable P2P VoD systems, and 2) streaming capacity for P2P VoD systems, respectively. Before we go deep into the resource optimization schemes, we will first present the overview of P2P streaming systems.

3.1 Overview of P2P Streaming Systems

With advances in broadband Internet access technology and the coding techniques, video streaming services have become increasingly popular. For example, YouTube [1], a video-sharing service which streams its videos to users on-demand, has attracted about 20 million views a day [2]. Video traffic is expected to be the dominating traffic on the Internet in the near future.

Conventionally, there are several approaches that can be used to deliver video streams to the users. We can roughly categorize the traditional approaches in two categories: unicast-based and multicast-based [3]. In unicast-based approaches, a unicast session is established for every user. There are three unicast approaches to deliver streaming video over the Internet: centralized, proxy, and Content Delivery Network (CDN) [3]. In the centralized approach, a powerful server with a high-bandwidth is deployed to serve video streams to the users. The server consumes a corresponding amount of bandwidth for each user. When the number of the users is increased to a certain level, the server may not be able to support them due to the bandwidth constraint. In the proxy approach, proxy servers are deployed near the client domains to cache a fraction of each video. The users can receive a part of the video from the proxy server near them. Therefore, this approach relieves the burden of the streaming server and the traffic load across the Wide Area Network (WAN). However, this approach requires deploying and managing proxies at many locations, which increases the cost. In the CDN approach, the video streams are delivered by the third party, known as the CDN. Content delivery networks, such as Akamai [4], deploy many servers at the edge of the Internet. The user can request the video from the most suitable server. The CDN effectively shortens the users' startup delays, and reduces the traffic across the WAN. However, the cost of the CDN approach is quite high since the CDN operator charges the video provider for every megabyte served.

The multicast-based approaches serve multiple users using the same stream, which is bandwidth efficient. The network-layer multicast establishes a tree over the internal routers with the users as the leaves of the tree. Though network-layer multicast is efficient, it is not widely deployed due to the following reasons. First, network-layer multicast requires routers to maintain a per-group state, which introduces high complexity at the IP layer. Second, it is difficult for the transport layer to provide flow control and congestion control in the network-layer multicast.

Peer-to-Peer (P2P) technology has recently been used in video streaming applications [18, 6]. The basic design philosophy of P2P is to encourage users to act as both clients and servers, namely as peers. In a P2P network, a peer not only downloads the data from other supplying peers who are buffering or storing them, but also uploads the downloaded data to other peers who are requesting them. The upload bandwidth of end users is efficiently utilized to offload the bandwidth burdens placed on the servers [7]. P2P streaming systems can be categorized into P2P live streaming systems and P2P VoD systems. A typical example of P2P live streaming systems is P2P IPTV, which is used to broadcast the TV programs to all the participating users. In a P2P VoD system, users can choose any video they like and start to watch it at any time. The playbacks of the same video on different users are not synchronized. In other words, each user watches a different position of the same video.

Next, we will present the overview of P2P live streaming systems and P2P VoD systems, respectively.

3.1.1 P2P Live Streaming Systems

Based on the overlay structure, P2P live streaming systems can be broadly classified into tree-based P2P live streaming systems and mesh-based P2P live streaming systems.

In a tree-Based P2P live streaming system, a single application-layer tree or multiple application-layer trees is constructed to deliver the video streams.

In a single-tree based P2P live streaming system, users participating in a live video streaming session can form a tree at the application layer. The root of the tree is the server. Each user joins the tree at a certain level. It receives the video from its parent peer at the level above and forwards the received video to its child peers at the level below. End System Multicast (ESM) [8] is a typical example of a single-tree based P2P live streaming system.

There are two major drawbacks for single-tree based P2P live streaming systems. First, the departure of a peer causes the isolation of all of its descendants from the video source. Second, not all the leaf nodes contribute their uploading bandwidths, which degrades the efficiency of the peer bandwidth utilization. A remedy to those drawbacks is a multiple-tree based P2P streaming system.

To improve the resiliency of the tree and the bandwidth utilization of the peers, multiple-tree based approaches have been proposed [9]. In multiple-tree based P2P live streaming systems, the video is encoded into multiple sub-streams, and each sub-stream is delivered over one tree. The quality received by a peer depends on the number of sub-streams that it receives. There are two key advantages to the multiple-tree solution. First, if a peer fails or leaves, all its descendants lose the sub-stream delivered from that peer, but they still receive the sub-streams delivered over the other trees. Therefore, all its descendants can receive a coarse video quality in case of a loss of a sub-stream. Second, a peer has different roles in different trees. It might be an internal node in one tree and a leaf node in another tree. When a peer is an internal node in a tree, its upload bandwidth will be utilized to upload the sub-stream delivered over that tree. To achieve high bandwidth utilization, a peer with a high upload bandwidth can supply sub-streams in more trees [10].

Peer dynamics make tree maintenance a challenging and costly task in tree-based P2P live streaming systems. To combat the peer dynamics, many recent P2P streaming systems use mesh-based streaming approach [18, 11]. In mesh-based P2P live streaming systems, each peer exchanges the data with a set of neighbors. If one neighbor leaves, the peer can still download the video data from the remaining neighbors. Meanwhile, the peers will add other peers into its neighbor set. Each peer can receive data from multiple supplying peers in mesh-based streaming systems, instead of a single parent in single-tree based streaming systems. Thus mesh-based streaming systems are robust against peer dynamics.

3.1.2   P2P VoD Systems

Different from live streaming systems in which each user watches almost the same position of the video, VoD service allows users to watch any point of the video at any time. VoD provides more flexibility and interactivity to users, thus attracting more users recently. In a VoD service, each user starts watching the video at a different time, and users may seek a new position in the video at any time. Therefore different users are watching different positions of the same video at a given instant. Due to the asynchrony of the users, the approaches of overlay construction in P2P VoD systems are different from those in P2P live streaming systems. In P2P live streaming systems, any peer can be a potential supplying peer (e.g., parent in tree-based systems or neighbor in mesh-based systems) of another peer since they are watching almost the same position. However, a peer in a P2P VoD system has to select a supplying peer from those candidates who have the requested content.

Depending on the forwarding approach, the existing P2P VoD systems can be classified into three categories: buffer-forwarding P2P VoD systems [9, 10], storage-forwarding P2P VoD systems [14, 12], and hybrid-forwarding P2P VoD systems [16]. In buffer-forwarding systems, each peer buffers the recently received content, and forwards it to the child peers. In storage-forwarding systems, the blocks of the video are stored in the storages of peers. When a peer wants to watch a video, it first looks for the supplying peers who are storing the content, and then requests the content from them. In hybrid-forwarding systems, a peer may have both buffer-forwarding parents and storage-forwarding parents, and may receive the video content from both kinds of parents.

In buffer-forwarding systems, if each peer is connected to multiple parents, a mesh-based overlay is formed. The maximal supported streaming rate in mesh-based architecture is improved compared to the tree-based architecture, since the upload bandwidth of each peer is utilized in a better way. In addition, the mesh-based overlay is more resilient to peer dynamics since each peer can receive the content supplied from multiple sources. However, each peer needs to address content reconciliation in a mesh-based overlay, which introduces extra communication overhead.

In buffer-forwarding P2P VoD systems, a peer receives the video content from its parents. If the parents jump to other positions in the video, the peer needs to search for new parents again. In storage-forwarding P2P VoD systems, each video is divided into multiple segments, which are distributed among peers. A peer receives the video content from the storage peers who possess the requested segment. Each storage peer would not change the video segments that it stores during its lifetime, and hence its activities (i.e., jumping) do not affect its children.

In buffer-forwarding P2P VoD systems, the peers redistribute their buffered content to their child peers, thus offloading the server burden. However, buffer-forwarding systems are less resilient to peer dynamics. If the parents of a peer perform random seeks and jump away from the current position, this peer needs to search for the new parents. In storage-forwarding P2P VoD systems, the storage peers do not change the stored segments, thus providing a table service. However, the achievable throughput in storage-forwarding systems is low due to the limitation of the total upload capacity. To overcome these drawbacks, a hybrid-forwarding P2P VoD system is proposed in [16], which improves the throughput by integrating both the buffer-forwarding approach and the storage-forwarding approach. In a hybrid-forwarding system, the peers watching the same video form an overlay using the buffer-forwarding approach. Moreover, each peer is encouraged to replicate in its storage one or multiple segments of the video that it has watched before. The stored segments can be used to serve other peers. The throughput of the hybrid-forwarding P2P VoD systems is improved by fully utilizing both the buffer resources and the storage resources of the peers.

3.2 Distributed Throughput Maximization for Scalable P2P VoD systems

Most existing P2P VoD systems adopt single-layer video coding. If the source rate is high, the peers with a limited or low download bandwidth may not be able to accommodate it. On the other hand, if the source rate is too low, the peers with a high download bandwidth may underutilize their download bandwidth. Therefore, a scalable source coding is a good solution for P2P VoD applications with heterogeneous bandwidth. With scalable coding, the high-bandwidth peers can receive a higher throughput, thus reconstructing the video at a higher quality, while the low-bandwidth peers can receive a lower bit rate, reconstructing the video at a lower quality. P2P streaming systems using scalable coding, such as layered coding [25] and Multiple Description Coding (MDC) [12], have been recently proposed.

In a scalable P2P VoD system, one of the goals is to maximize the aggregate throughput among all the peers. In the P2P networks, each peer may receive streams from multiple suppliers. Therefore, there is a risk that the receiver may receive redundant packets. To combat this delivery redundancy, the receiver can negotiate with its suppliers and request the distinct packets. However, this method will introduce extra message exchanges among peers. Alternatively the coding techniques such as fountain codes [18][19] or network coding [4] can be applied to eliminate the redundant packets. If each peer receives distinct packets, maximizing the reconstructed quality at each peer is equivalent to maximizing the throughput.

The aggregate throughput can be maximized through optimal link rate allocation. However, it is quite challenging to allocate each link rate optimally. The reasons are as follows. 1) The number of link rates is typically huge in P2P applications. 2) There are correlations among different link rates. For example, the outgoing rate from a peer can not exceed the aggregate rate received by this peer. 3) Each peer has only the local information from its neighbors rather than the global information. Hence, a centralized allocation algorithm is not appropriate for P2P applications.

Throughput maximization in P2P applications has not been systematically studied in the literature. Some existing P2P systems use an equal allocation scheme by equally allocating the upload bandwidth to each of its outgoing links [21]. Some other systems use a proportional allocation scheme [2], in which the rate allocated to a child peer is proportional to the available download bandwidth of this child peer. These existing approaches suffer from inefficient usage of network resources in a large and heterogeneous environment, thus they cannot achieve a maximal throughput.

In this section, we present a systematic study on the throughput maximization in P2P VoD applications. First, we formulate the throughput maximization problem in the existing buffer-forwarding P2P VoD system, and develop a fully distributed algorithm to solve it. Second, we propose a hybrid-forwarding P2P VoD architecture, which improves the aggregate throughput greatly, compared to the existing buffer-forwarding architecture. We formulate the throughput maximization problem in the hybrid-forwarding P2P VoD architecture, and solve the problem with a fully distributed algorithm.

3.2.1 Throughput Maximization in Buffer-Forwarding P2P VoD Systems

In this subsection, we examine the throughput maximization problem in buffer-forwarding P2P VoD systems. We will present problem formulation, distributed solution, and simulation results, respectively.

3.2.1.1 Problem Formulation

In buffer-forwarding P2P VoD systems, the server and the peers watching the same video organize themselves into an overlay, which can be modeled as a directed graph $\mathbf{G} = (\mathbf{N}, \mathbf{L})$, where $\mathbf{N}$ is the set of peers and $\mathbf{L}$ is the set of directed overlay links. Peer 1 is defined as the server.

In order to formulate the throughput maximization problem, we first define the matrices to represent the topology of the overlay. The relationship between the node and its outgoing links can be represented by a matrix $\mathbf{A}$, whose elements are defined by

$$a_{il} = \begin{cases} 1, & \text{if link } l \text{ is an outgoing link from node } i, \\ 0, & \text{otherwise.} \end{cases} \tag{3.1}$$

The relationship between the node and its incoming links is represented by a matrix $\mathbf{B}$, whose elements are given by

$$b_{il} = \begin{cases} 1, & \text{if link } l \text{ is an incoming link into node } i, \\ 0, & \text{otherwise.} \end{cases} \tag{3.2}$$

The relationship between the outgoing link and the incoming link of a node is represented by a matrix $\mathbf{C}$, whose elements are given by

$$c_{lm} = \begin{cases} 1, & \text{if link } m \text{ is an incoming link into the start} \\ & \text{node of link } l, \\ 0, & \text{otherwise.} \end{cases} \tag{3.3}$$

In VoD streaming applications, in-time delivery of the packets is important. The packets that cannot arrive at the receiver before the payback deadline are useless for decoding. In buffer-forwarding VoD systems, each peer caches the received packets in the buffer to smoothen the payback and serve other peers. Fig. 2 shows the buffer at a peer. The start time of the buffer at peer j is denoted as t_j^{bs}, the end time of the buffer at peer j is denoted as t_j^{be}, and the playback time at peer j is denoted as t_j^p. Suppose that a serving peer, peer

j, transmits the buffered packets to a receiving peer, peer i, over an overlay link, link l. With an initial playback delay d_i^{in}, peer i requests the packets with timestamps larger than $(t_i^p + d_i^{in})$. Therefore, only the packets within a *transmission sliding window* at peer j are eligible for transmission to peer i. In order to be a serving peer to peer i, peer j needs to 1) have a larger playback time than peer i, and 2) have a buffer overlap with peer i. If $t_j^{bs} < t_i^p + d_i^{in}$, the start time of the transmission sliding window is given by $t_{ji}^s = t_i^p + d_i^{in}$, as shown in Fig. 3.1(a). If $t_j^{bs} \geq t_i^p + d_i^{in}$, the start time of the transmission sliding window is given by $t_{ji}^s = t_j^{bs}$, as shown in Fig. 3.1(b). In other words, $t_{ji}^s = \max\{t_j^{bs}, t_i^p + d_i^{in}\}$. The end time of the transmission sliding window is $t_{ji}^e = t_i^{be}$. Then the duration of the sliding window is given by $d_{sw} = t_{ji}^e - t_{ji}^s$. Suppose that the transmission delay over link l from peer j to peer i follows an exponential distribution with a mean delay τ_l. The packets need to arrive at peer i before the payback deadline, otherwise they will be discarded. For a delay upper bound d_{up}, the probability that a packet transmitted to peer i over link l is discarded is given by

$$p_d(d_{up}) = \int_{d_{up}}^{\infty} \frac{1}{\tau_l} \exp(-\frac{t}{\tau_l}) dt = \exp(-\frac{d_{up}}{\tau_l}). \tag{3.4}$$

Any packet falling within the transmission sliding window at peer j has a probability to be transmitted at the current moment. Which packet will be chosen for transmission dynamically depends on: 1) which packets are available at peer i, and 2) which packets are requested from other concurrent serving peers. We assume that the packets within the transmission sliding window at peer j have equal probability to be scheduled for transmission. Then the expected probability that the transmitted packets arrive after the playback deadline at peer i is given by

$$\begin{aligned} p_l &= \int_{t_{ji}^s - t_i^p}^{t_{ji}^e - t_i^p} \frac{p_d(d_{up})}{d_{sw}} d(d_{up}) \\ &= \int_{t_{ji}^s - t_i^p}^{t_{ji}^e - t_i^p} \frac{1}{d_{sw}} \exp(-\frac{d_{up}}{\tau_l}) d(d_{up}) \\ &= \frac{\tau_l}{d_{sw}} (\exp(-\frac{t_{ji}^s - t_i^p}{\tau_l}) - \exp(-\frac{t_{ji}^e - t_i^p}{\tau_l})), \quad \forall l \in \mathbf{L}. \end{aligned} \tag{3.5}$$

The throughput at the receiving peer is defined as the useful received rate, excluding the excessively delayed packets which do not catch the playback deadline. Therefore the throughput at peer i can be denoted as $\sum_{l \in \mathbf{L}} b_{il}(x_l(1 - p_l))$, where x_l is the link rate at link l. In practical VoD applications, we can provide a differentiated video quality to a user based on the service the user purchases. For example, the users who pay a higher price for the on-demand video should receive a higher throughput. To enable a differentiated throughput, we allocate a priority weight χ_i to peer i. Our objective is to maximize the aggregate weighted throughput, which is formulated by $\sum_{i \in \mathbf{N}} \chi_i \sum_{l \in \mathbf{L}} b_{il}(x_l(1 - p_l))$.

At the server, each video is layered encoded and packetized in a prioritized

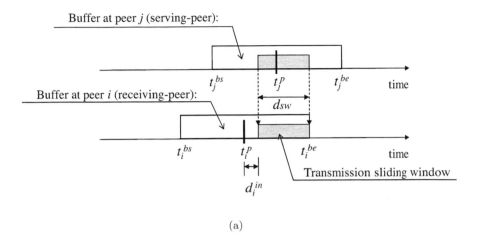

(a)

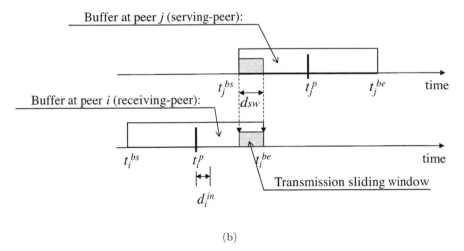

(b)

FIGURE 3.1
Transmission sliding window for the packet transmission over the overlay link
l from peer j to peer i: (a) if $t_j^{bs} < t_i^p + d_i^{in}$, and (b) if $t_j^{bs} \geq t_i^p + d_i^{in}$

way before it is transmitted. The maximal source rate of the video is denoted by s_r. If s_r is small, the upload bandwidth in the system will be underutilized. Therefore, we encode the video into a high s_r. Any peer can at most receive a rate of s_r. This is regarded as a *source rate constraint*, which is formulated mathematically as $\sum_{l \in \mathbf{L}} b_{il} x_l \le s_r, \forall i \in \mathbf{N}$. If a peer receives a rate of s_r, it can reconstruct the video at a full quality. Otherwise, it will reconstruct a video at a partial quality if the received rate is less than s_r.

Peer i has a download capacity (denoted as I_i) and an upload capacity (denoted as O_i). In P2P streaming, the bandwidth bottleneck usually occurs at the access links. Therefore, peer i has a *download bandwidth constraint*, formulated as $\sum_{l \in \mathbf{L}} b_{il} x_l \le I_i, \forall i \in \mathbf{N}$, and an *upload bandwidth constraint*, formulated as $\sum_{l \in \mathbf{L}} a_{il} x_l \le O_i, \forall i \in \mathbf{N}$, respectively.

At peer i ($i \ne 1$), each outgoing link from peer i carries a rate no larger than the total incoming rate into that peer. Each direct outgoing link from the server carries a rate no larger than s_r. This constraint is referred to as a *link-forwarding constraint*, which can be formulated as

$$x_l - \sum_{m \in \mathbf{L}} c_{lm} x_m \le \sigma_l, \quad \forall l \in \mathbf{L}. \tag{3.6}$$

where σ_l is a link-forwarding compensation element, and it is defined as

$$\sigma_l = \begin{cases} s_r, & \text{if link } l \text{ is a direct outgoing link from the server,} \\ 0, & \text{otherwise.} \end{cases} \tag{3.7}$$

The throughput maximization problem in the buffer-forwarding P2P VoD system is to maximize the aggregate throughput by optimally allocating the link rates, subject to the four constraints: *the source rate constraint, the download bandwidth constraint, the upload bandwidth constraint* and *the link-forwarding constraint*. Mathematically, the optimization problem can be formulated as follows.

$$\begin{aligned} \text{maximize}_{(\mathbf{x})} \quad & \sum_{i \in \mathbf{N}} \chi_i \sum_{l \in \mathbf{L}} b_{il}(x_l(1 - p_l)) \\ \text{subject to} \quad & \sum_{l \in \mathbf{L}} b_{il} x_l \le s_r, \quad \forall i \in \mathbf{N}, \\ & \sum_{l \in \mathbf{L}} b_{il} x_l \le I_i, \quad \forall i \in \mathbf{N}, \\ & \sum_{l \in \mathbf{L}} a_{il} x_l \le O_i, \quad \forall i \in \mathbf{N}, \\ & x_l - \sum_{m \in \mathbf{L}} c_{lm} x_m \le \sigma_l, \quad \forall l \in \mathbf{L}, \\ & p_l = \frac{\tau_l}{d_{sw}}\left(\exp(-\frac{t_{ji}^s - t_i^p}{\tau_l}) - \exp(-\frac{t_{ji}^e - t_i^p}{\tau_l})\right), \\ & \hspace{6cm} \forall l \in \mathbf{L}, \\ & x_l \ge 0, \quad \forall l \in \mathbf{L}. \end{aligned} \tag{3.8}$$

The problem in (3.8) is a Linear Programming (**LP**) problem [21]. It can be solved in a centralized way using the simplex method or the interior point method [21]. However, a centralized algorithm requires global information, thus is not applicable for the distributed P2P networks. In order to develop a distributed algorithm, we change the maximization problem (3.8) to an equivalent minimization problem, and then revise the objective function by adding a corresponding quadratic regularization term for each link rate variable. Then the optimization problem is converted to:

$$
\begin{aligned}
\text{minimize}_{(\mathbf{x})} \quad & -\sum_{i\in\mathbf{N}} \chi_i \sum_{l\in\mathbf{L}} b_{il}(x_l(1-p_l)) + \varepsilon\Sigma_{l\in\mathbf{L}} x_l^2 \\
\text{subject to} \quad & \sum_{l\in\mathbf{L}} b_{il}x_l \le y_i, \quad \forall i \in \mathbf{N}, \\
& \sum_{l\in\mathbf{L}} a_{il}x_l \le O_i, \quad \forall i \in \mathbf{N}, \\
& x_l - \sum_{m\in\mathbf{L}} c_{lm}x_m \le \sigma_l, \quad \forall l \in \mathbf{L}, \\
& x_l \ge 0, \quad \forall l \in \mathbf{L}, \\
& y_i = min\{I_i, s_r\}, \quad \forall i \in \mathbf{N},
\end{aligned}
\tag{3.9}
$$

where $\varepsilon(\varepsilon > 0)$ is called the *regularization factor*. When ε is small enough, the solution for the problem (3.9) is arbitrarily close to the solution for the original throughput maximization problem (3.8).

3.2.1.2 Distributed Solution

In the optimization problem (3.9), the objective function is strictly convex, and the constraints are linear. Hence, it is a convex optimization problem [19]. We can develop a fully distributed algorithm to solve the optimization problem (3.9) using the Lagrangian duality properties.

Slater's condition holds [19] for the convex optimization problem (3.9), thus the duality gap is zero. We can obtain the primal optimal solutions indirectly by first solving the dual problem [20]. We introduce dual variables $u_i, v_i(\forall i \in \mathbf{N})$ and $\lambda_l(\forall l \in \mathbf{L})$ to formulate the Lagrangian corresponding to the primal problem (3.9) as below

$$
\begin{aligned}
& L(\mathbf{x}, \mathbf{u}, \mathbf{v}, \lambda) \\
& = \varepsilon\sum_{l\in\mathbf{L}} x_l^2 - \sum_{i\in\mathbf{N}} \chi_i \sum_{l\in\mathbf{L}} b_{il}(x_l(1-p_l)) \\
& + \sum_{i\in\mathbf{N}} u_i(\sum_{l\in\mathbf{L}} b_{il}x_l - y_i) + \sum_{i\in\mathbf{N}} v_i(\sum_{l\in\mathbf{L}} a_{il}x_l - O_i) \\
& + \sum_{l\in\mathbf{L}} \lambda_l(x_l - \sum_{m\in\mathbf{L}} c_{lm}x_m - \sigma_l) \\
& = \sum_{l\in\mathbf{L}}(\varepsilon x_l^2 + x_l q_l) - \sum_{l\in\mathbf{L}} \lambda_l\sigma_l - \sum_{i\in\mathbf{N}}(u_iy_i + v_iO_i),
\end{aligned}
\tag{3.10}
$$

where $q_l = -(1-p_l)\sum_{i\in\mathbf{N}} \chi_i b_{il} + \sum_{i\in\mathbf{N}} b_{il}u_i + \sum_{i\in\mathbf{N}} a_{il}v_i + \lambda_l - \sum_{m\in\mathbf{L}} \lambda_m c_{ml}$. The Lagrange dual function is the minimum value of the Lagrangian over the link rates:

$$
\begin{aligned}
& G(\mathbf{u}, \mathbf{v}, \lambda) \\
& = min_{(\mathbf{x}\ge 0)}\{L(\mathbf{x}, \mathbf{u}, \mathbf{v}, \lambda)\} \\
& = \sum_{l\in\mathbf{L}} min_{(x_l\ge 0)}(\varepsilon x_l^2 + x_l q_l) - \sum_{l\in\mathbf{L}} \lambda_l\sigma_l - \\
& \quad \sum_{i\in\mathbf{N}}(u_iy_i + v_iO_i).
\end{aligned}
\tag{3.11}
$$

The Lagrange dual problem is to maximize the Lagrange dual function. That is:

$$
\begin{aligned}
\text{maximize}_{(\mathbf{u},\mathbf{v},\lambda)} \quad & G(\mathbf{u}, \mathbf{v}, \lambda) \\
\text{subject to} \quad & u_i \ge 0, \quad \forall i \in \mathbf{N}, \\
& v_i \ge 0, \quad \forall i \in \mathbf{N}, \\
& \lambda_l \ge 0, \quad \forall l \in \mathbf{L}.
\end{aligned}
\tag{3.12}
$$

We use the subgradient method [21] to solve the Lagrange dual problem

(3.12). The dual variables $u_i^{(k+1)}, v_i^{(k+1)}, \lambda_l^{(k+1)}$ at the $(k+1)^{th}$ iteration are updated respectively by

$$u_i^{(k+1)} = \max\{0, u_i^{(k)} - \theta^{(k)}(y_i - \sum_{l \in \mathbf{L}} b_{il} x_l^{(k)})\}, \quad \forall i \in \mathbf{N}, \tag{3.13}$$

$$v_i^{(k+1)} = \max\{0, v_i^{(k)} - \theta^{(k)}(O_i - \sum_{l \in \mathbf{L}} a_{il} x_l^{(k)})\}, \quad \forall i \in \mathbf{N}, \tag{3.14}$$

$$\lambda_l^{(k+1)} = \max\{0, \lambda_l^{(k)} - \theta^{(k)}(\sum_{m \in \mathbf{L}} c_{lm} x_m^{(k)} + \sigma_l - x_l^{(k)})\}, \quad \forall l \in \mathbf{L}, \tag{3.15}$$

where $\theta^{(k)} > 0$ is the step size at k^{th} iteration. The algorithm is guaranteed to converge to the optimal value for a step-size sequence satisfying the non-summable diminishing rule [20]:

$$\lim_{k \to \infty} \theta^{(k)} = 0, \quad \sum_{k=1}^{\infty} \theta^{(k)} = \infty. \tag{3.16}$$

In our algorithm, the step size at the kth iteration is given by: $\theta^{(k)} = \omega/\sqrt{k}$, where $\omega > 0$.

We can compute the link rates in parallel. At the kth iteration, the link rate $x_l^{(k)}$ at link l can be calculated from the dual variables.

$$x_l^{(k)} = \max\{0, \frac{-q_l^{(k)}}{2\varepsilon}\}, \quad \forall l \in \mathbf{L}, \tag{3.17}$$

where $q_l^{(k)} = -(1 - p_l)\sum_{i \in \mathbf{N}} \chi_i b_{il} + \sum_{i \in \mathbf{N}} b_{il} u_i^{(k)} + \sum_{i \in \mathbf{N}} a_{il} v_i^{(k)} + \lambda_l^{(k)} - \sum_{m \in \mathbf{L}} \lambda_m^{(k)} c_{ml}$.

In practical distributed protocol, peer i maintains the following variables: 1) the incoming and outgoing link rates, 2) the dual variables u_i and v_i, and 3) the dual variable λ_l, where l is an outgoing link from peer i. For example, peer i in Fig. 3.2 maintains the link rates x_a, x_b, x_c, x_d and dual variables $u_i, v_i, \lambda_c, \lambda_d$. At each iteration, peer i performs two computation tasks: 1) update the dual variables maintained by it, and 2) update the outgoing link rates. Peer i updates the associated dual variables using Equations (3.13), (3.14) and (3.15), which require only local link rates. Therefore there is no communication overhead for the update of the associated dual variables. Peer i updates the rate of each outgoing link using Equation (6.16), which requires the information from the downstream node of this link. Fig. 3.2 shows the message transmission from peer k to peer i when peer i wants to update one of its outgoing link rates x_c. Based on Equation (6.16), we know that peer i needs to have $(v_i, u_k, \lambda_c, \lambda_e, \lambda_f)$ in order to compute x_c. The variables v_i and λ_c are locally available at peer i, while the variables $(u_k, \lambda_e, \lambda_f)$ are maintained by peer k. Therefore a message containing $(u_k, \lambda_e, \lambda_f)$ needs to be delivered to peer i for the update of x_c. The communication overhead for the update of a link rate is just a message from the downstream node of the link. The message contains only a few numbers, thus consuming a small bandwidth.

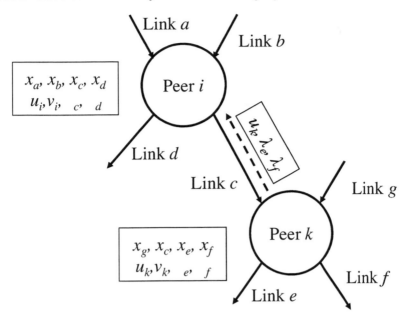

FIGURE 3.2
Message transmission from peer k to peer i

3.2.1.3   Simulation Results

We perform extensive simulations to study the throughput maximization in buffer-forwarding P2P VoD systems. The simulation setup is as follows. The server has an upload capacity of 2 Mbps. The maximal source rate is 1.3 Mbps. Peers are heterogeneous at upload and download capacity. There are two classes of peers: cable/DSL peers and Ethernet peers. Cable/DSL peers take 85% of the total population, with download capacity uniformly distributed between 0.6 Mbps and 1.0 Mbps and upload capacity uniformly distributed between 0.2 Mbps and 0.4 Mbps. Ethernet peers take the remaining 15% of the total population, with both upload and download capacities uniformly distributed between 1.0 Mbps and 2.0 Mbps. The playback time of each peer is uniformly distributed between 0 and the length of the video (40 minutes). In the overlay construction, each receiving peer is connected to two buffer-forwarding parents. The transmission delay of an overlay link is uniformly distributed between 5.0 ms and 20.0 ms. We encode Foreman sequence in Common Intermediate Format (CIF) into 8 layers using SNR-scalable extension of H.264/AVC [14]. Each GOP consists of 16 frames. Each layer has an average source bit rate of 60.0 Kbps. The source bit stream is packetized into 8 descriptions, each of which has an average bit rate of 162.5 Kbps. In the distributed optimization, the step size at k^{th} iteration is given by $\theta^{(k)} = 1.3/\sqrt{k}$, and the regularization factor is set to 0.4. The convergence threshold for the dual function is set to 10^{-5}.

According to the measurement study from the real P2P VoD system [28] deployed in 2006, the number of the concurrent users watching the same video at the peak time can reach several hundreds. Therefore, we vary the network size from 100 to 500 peers, and compare the average throughput and the Peak Signal-to-Noise Ratio (PSNR), respectively. Fig. 3.3(a) compares the average throughput among different allocation schemes: a) *centralized LP*, in which the link rates are obtained by solving the optimization problem (3.8) using the centralized algorithm; b) *distributed optimization*, in which the link rates are obtained by solving the optimization problem (3.9) using the proposed distributed algorithm; c) *proportional allocation scheme*, in which the link rate allocated to a child node is proportional to the available download bandwidth of this child node; and d) *equal allocation scheme*, in which each peer allocates the upload rate equally to its children. Compared to the centralized LP solution, the proposed distributed algorithm sacrifices about 1.0% of the average throughput due to the introduction of the small regularization terms. However, the centralized LP algorithm requires global information, which is not appropriate for distributed P2P applications. The proposed distributed algorithm achieves on average 10.48% of the throughput improvement compared to the proportional allocation scheme, and on average 17.70% of the throughput improvement compared to the equal allocation scheme. The comparison of the average reconstructed PSNR is depicted in Fig. 3.3(b). By applying network coding, each peer receives distinct packets. A higher throughput leads to a higher PSNR. Therefore we observe that the proposed scheme using distributed optimization achieves an average PSNR improvement of 0.30 dB compared to the proportional allocation scheme, and 0.55 dB compared to the equal allocation scheme.

When the number of the peers is increased from 100 to 500, the number of the overlay links is increased linearly from 197 to 997, as shown in Fig. 3.4(a). In the proposed algorithm, a link rate is updated only when the corresponding dual variables have been changed. From Fig. 3.4(b), we can see that the average number of iterations per link basically remains at the same level. The communication overhead varies little with different network size, as shown in Fig. 3.4(c). The number of iterations per link and the communication overhead per node do not increase with the number of the peers, demonstrating that the proposed algorithm is scalable. In the proposed algorithm, the update of a link rate requires only a message delivery from the corresponding downstream node. Therefore, the average communication overhead per node is very small, consuming less than 1.0% of the average throughput.

There is a tradeoff between the throughput and the communication overhead with different regularization factors, as shown in Fig. 3.5. If a larger regularization factor is chosen, the communication overhead can be reduced, On the other hand, a larger regularization factor leads to a reduction of the average throughput. In our simulations, we choose a regularization factor of

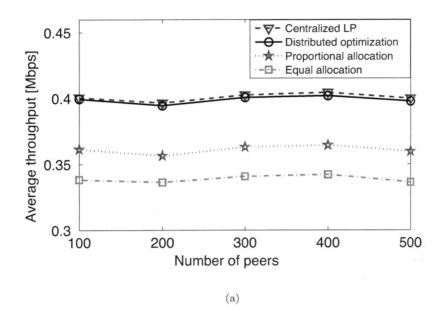

(a)

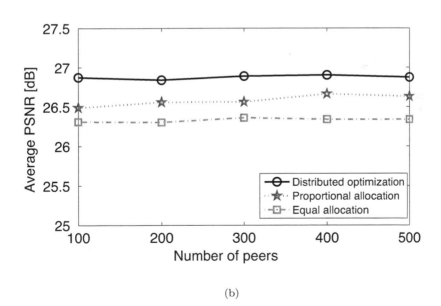

(b)

FIGURE 3.3
Performance comparison in buffer-forwarding P2P VoD systems with different network sizes: (a) the average throughput, and (b) the average PSNR

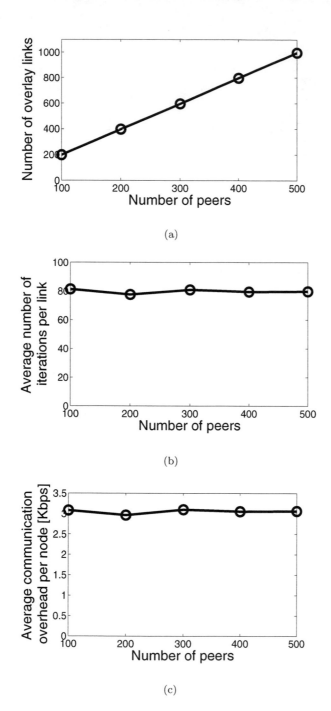

(a)

(b)

(c)

FIGURE 3.4
Comparison of the cost introduced by the proposed distributed algorithm
in buffer-forwarding P2P VoD systems with different network sizes: (a) the
number of the overlay links, (b) the average number of iterations per link,
and (c) the average communication overhead per node

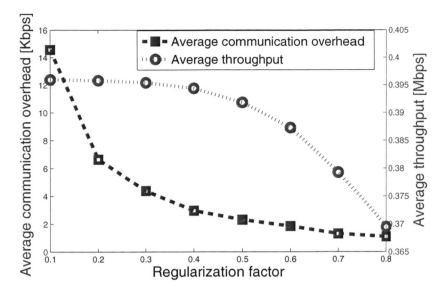

FIGURE 3.5

The impact of the regularization factor on the communication overhead and the average throughput in a 200-peer buffer-forwarding P2P VoD system

0.4, with which we obtain an average throughput of 0.394 Mbps at the average communication overhead per node of 2.96 Kbps.

3.2.2 Throughput Maximization in Hybrid-Forwarding P2P VoD Systems

To improve the throughput, we propose a hybrid forwarding P2P VoD architecture which combines both the buffer-forwarding approach and the storage-forwarding approach. In the hybrid-forwarding architecture, the peers watching the same video form an overlay using the buffer-forwarding approach. In the example shown in Fig. 3.6, peers 3, 7, 130, 116, 104 and 1 form video-1 buffer-forwarding overlay, and peers 134, 34, 115, 43, 22 and 1 form video-2 buffer-forwarding overlay. Please note that the server participates in all of the video overlays since it contains the contents of all the videos. In the hybrid-forwarding architecture, each peer is encouraged or required to replicate in its storage one or multiple segments of the video that it has watched before. The stored segments can be used to serve other peers. Therefore, the upload bandwidth of the idle peers can be utilized, since they can contribute their stored segments to other peers. In the hybrid-forwarding architecture, peers may have both buffer-forwarding links and storage-forwarding links. For example, peer 115 in Fig. 3.6 forwards its buffered content to peer 34 and peer 134 in video-2 overlay; meanwhile, it also forwards its stored content to peer 7

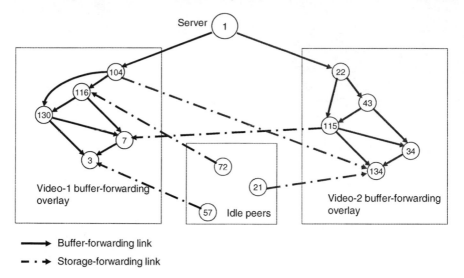

FIGURE 3.6
Illustration of a hybrid-forwarding P2P VoD system

in the video-1 overlay. The two kinds of outgoing links compete for the limited upload bandwidth. By allocating a larger portion of the upload bandwidth to the peers watching a high prioritized video (e.g., paid program) compared to those watching a low prioritized video (e.g., free program), service differentiation among videos can be implemented. Furthermore, the hybrid-forwarding architecture is robust to random seeks because the stored content is a stable source unless the supplier shuts down the P2P application. For example, when the buffer-forwarding parents (e.g., peer 130 and 116) of peer 7 both jump to other positions, peer 7 still has a stream supply from its storage-forwarding parent (e.g., peer 115), which maintains the continuous playback.

In the overlay construction, the peers watching the same video are first organized into a buffer-forwarding overlay using one of the existing approaches in [7][8][9], then each peer locates the storage-forwarding parents either in a centralized way [12] or in a distributed way [11] and connects itself to them.

3.2.2.1   Problem Formulation

The hybrid-forwarding overlay can be represented by a directed graph $\mathbf{G} = (\mathbf{N}, \mathbf{L})$, where $\mathbf{N}$ is the set of the peers and $\mathbf{L}$ is the set of directed overlay links. The relationship between the node and its incoming links is represented by a matrix $\mathbf{B}$, which is defined in Equation (3.2) in Section 3.2.1.1. The relationship between the outgoing link and the incoming link of a node is represented by a matrix $\mathbf{C}$, which is defined in Equation (3.3) in Section 3.2.1.1. The relationship between the node and its buffer-forwarding outgoing links is represented by a matrix $\mathbf{D}$, whose elements are given by

$$
d_{il} = \begin{cases} 1, & \text{if link } l \text{ is a buffer-forwarding outgoing link} \\ & \text{from node } i, \\ 0, & \text{otherwise.} \end{cases} \tag{3.18}
$$

The relationship between the node and its storage-forwarding outgoing links is represented by a matrix $\mathbf{H}$, whose elements are given by

$$
h_{il} = \begin{cases} 1, & \text{if link } l \text{ is a storage-forwarding outgoing link} \\ & \text{from node } i, \\ 0, & \text{otherwise.} \end{cases} \tag{3.19}
$$

Suppose there are V videos, denoted as $v = 1, ..., V$, being distributed in the P2P overlay. Each video is allocated with a priority level, denoted as $\alpha_v (0 \leq \alpha_v \leq 1)$. A peer can only watch one video at a time. We use β_i to denote the peer priority based on which video the peer is watching:

$$
\beta_i = \begin{cases} \alpha_v, & \text{if peer } i \text{ is watching video } v, \\ 0, & \text{otherwise.} \end{cases} \tag{3.20}
$$

The probability that a packet transmitted through link l arrives after the playback deadline is denoted by p_l, which can be computed using Equation (3.5) in Section 3.2.1.1. The throughput at peer i is given by $\sum_{l \in \mathbf{L}} b_{il}(x_l(1 - p_l)), \forall i \in \mathbf{N}$. The objective of the optimization problem is to maximize the aggregate weighted throughput, taking into account the video priority level. It is formulated by $\sum_{i \in \mathbf{N}} \beta_i \sum_{l \in \mathbf{L}} b_{il}(x_l(1 - p_l))$.

As described in Section 3.2.1.1, each peers follows the *source rate constraint*, which is given by $\sum_{l \in \mathbf{L}} b_{il} x_l \leq s_r, \forall i \in \mathbf{N}$, where s_r is the maximal source rate. Also, each peer follows the *download bandwidth constraint*: $\sum_{l \in \mathbf{L}} b_{il} x_l \leq I_i, \forall i \in \mathbf{N}$. The aggregate rate of the buffer-forwarding outgoing links from peer i can be expressed by $\sum_{l \in \mathbf{L}} d_{il} x_l$, and the aggregate rate of the storage-forwarding outgoing links from peer i can be expressed by $\sum_{l \in \mathbf{L}} h_{il} x_l$. The total outgoing rate, consisting of the buffer-forwarding outgoing rate and the storage-forwarding outgoing rate, needs to satisfy the *upload bandwidth constraint*, given by $\sum_{l \in \mathbf{L}} d_{il} x_l + \sum_{l \in \mathbf{L}} h_{il} x_l \leq O_i, \forall i \in \mathbf{N}$.

The direct outgoing link from the server carries a rate no larger than s_r, and each buffer-forwarding outgoing link from peer $i (i \neq 1)$ carries a rate no larger than the total incoming rate into that peer. This constraint is referred to as a *buffer-forwarding constraint*, which can be formulated as

$$
x_l - \sum_{m \in \mathbf{L}} c_{lm} x_m \leq \varphi_l, \quad \forall l \in \mathbf{L}. \tag{3.21}
$$

where φ_l is a buffer-forwarding compensation element, and it is defined as

$$
\varphi_l = \begin{cases} s_r, & \text{if link } l \text{ is a direct outgoing link from the} \\ & \text{server or a storage-forwarding link,} \\ 0, & \text{otherwise.} \end{cases} \tag{3.22}
$$

How to replicate the segments in the pool of the peers is still an open problem. In this paper, we adopt a random replication strategy in which each peer is required to store one random segment. Our algorithm can be easily extended to deal with the case in which multiple segments are replicated in each peer. The stored segments are obtained in the previous VoD sessions. Since each peer is allocated with a different priority weight, it receives the video at a different rate. The peer with a high priority weight receives the video at a high rate, while the peer with a low priority weight receives the video at a low rate. Therefore, each peer has a different rate for the stored segment. We use $F_i (\forall i \in \mathbf{N})$ to denote the rate of the stored segment at peer i. In the hybrid-forwarding P2P VoD system, we have a *storage-forwarding constraint*: the rate at each storage-forwarding outgoing link from peer i is no larger than the rate F_i of the stored segment at peer i. This constraint can be formulated as

$$x_l - \sum_{i \in \mathbf{N}} h_{il} F_i \leq \xi_l, \quad \forall l \in \mathbf{L}. \tag{3.23}$$

where ξ_l is a storage-forwarding compensation element, and it is defined as

$$\xi_l = \begin{cases} s_r, & \text{if link } l \text{ is a buffer-forwarding link,} \\ 0, & \text{otherwise.} \end{cases} \tag{3.24}$$

The throughput maximization problem in the hybrid-forwarding P2P VoD system is to maximize the aggregate weighted throughput by optimally determining the link rates, subject to the above five constraints: *source rate constraint, download bandwidth constraint, upload bandwidth constraint, buffer-forwarding constraint*, and *storage-forwarding constraint*. Mathematically, the optimization problem can be formulated as follows.

$$
\begin{array}{ll}
\text{maximize}_{(\mathbf{x})} & \sum_{i \in \mathbf{N}} \beta_i \sum_{l \in \mathbf{L}} b_{il}(x_l(1 - p_l)) \\
\text{subject to} & \sum_{l \in \mathbf{L}} b_{il} x_l \leq s_r, \qquad \forall i \in \mathbf{N}, \\
& \sum_{l \in \mathbf{L}} b_{il} x_l \leq I_i, \qquad \forall i \in \mathbf{N}, \\
& \sum_{l \in \mathbf{L}} d_{il} x_l + \sum_{l \in \mathbf{L}} h_{il} x_l \leq O_i, \quad \forall i \in \mathbf{N}, \\
& x_l - \sum_{m \in \mathbf{L}} c_{lm} x_m \leq \varphi_l, \qquad \forall l \in \mathbf{L}, \\
& x_l - \sum_{i \in \mathbf{N}} h_{il} F_i \leq \xi_l, \qquad \forall l \in \mathbf{L}, \\
& x_l \geq 0, \qquad \forall l \in \mathbf{L}.
\end{array} \tag{3.25}
$$

The optimization problem (3.25) is a **LP** problem. Similar to the conversion in Section 3.2.1.1, we change the objective function in the optimization problem (3.25) to a strictly convex function. Then the optimization problem is converted to:

$$
\begin{array}{ll}
\text{minimize}_{(\mathbf{x})} & -\sum_{i \in \mathbf{N}} \beta_i \sum_{l \in \mathbf{L}} b_{il}(x_l(1 - p_l)) + \varepsilon \sum_{l \in \mathbf{L}} x_l^2 \\
\text{subject to} & \sum_{l \in \mathbf{L}} b_{il} x_l \leq y_i, \quad \forall i \in \mathbf{N}, \\
& \sum_{l \in \mathbf{L}} d_{il} x_l + \sum_{l \in \mathbf{L}} h_{il} x_l \leq O_i, \quad \forall i \in \mathbf{N}, \\
& x_l - \sum_{m \in \mathbf{L}} c_{lm} x_m \leq \varphi_l, \quad \forall l \in \mathbf{L}, \\
& x_l - \sum_{i \in \mathbf{N}} h_{il} F_i \leq \xi_l, \quad \forall l \in \mathbf{L}, \\
& x_l \geq 0, \quad \forall l \in \mathbf{L}, \\
& y_i = \min\{I_i, s_r\}, \quad \forall i \in \mathbf{N}.
\end{array} \tag{3.26}
$$

A sufficiently small regularization factor $\varepsilon(\varepsilon > 0)$ makes the solution for the optimization problem (3.26) arbitrarily close to the solution for the original problem (3.25).

The optimization problem (3.26) has a quadratic convex objective function and the linear constraints, thus it is a convex optimization problem [19]. We can develop a fully distributed algorithm to solve it [20].

3.2.2.2 Simulation Results

In the simulations for throughput maximization in hybrid-forwarding P2P VoD systems, we use the same setup as described in Section 3.2.1.3. Two videos, denoted as video 1 and video 2, are distributed over the hybrid-forwarding P2P VoD system. The server owns the whole length of the two videos at the full rate in its storage. Among all the peers, 60% percent of the peers are watching video 1, 30% of the peers are watching video 2, and the remaining 10% of the peers are idle at that moment. The priority levels of two videos are both set to a default value 0.5 if not specified in particular. Each peer is required to store a random segment with a length of 2 minutes. In the default setting, each receiver is connected to two buffer-forwarding parents, and two storage-forwarding parents if available.

We compare the performance with different network sizes. "Hybrid-forwarding centralized" represents the results obtained by solving the optimization problem (3.25) with the centralized algorithm in the hybrid-forwarding architecture. "*Hybrid-forwarding distributed*" represents the results obtained by solving the optimization problem (3.26) with the proposed distributed algorithm in the hybrid-forwarding architecture. "*Buffer-forwarding distributed*" represents the solution obtained by a distributed algorithm in the buffer-forwarding architecture. "*Buffer-forwarding proportional*" represents the results obtained by using proportional allocation scheme in the buffer-forwarding architecture. "*Storage-forwarding distributed*" represents the solution obtained by a distributed algorithm in the storage-forwarding architecture. "*Hybrid-forwarding centralized*" obtains the truly optimal solution in the hybrid-forwarding architecture. However, this scheme requires global information, thus not scalable. As shown in Fig. 3.7(a), the average throughput obtained with the proposed algorithm, "*hybrid-forwarding distributed*," is 6.6% – 9.0% lower than the truly optimal throughput due to the introduction of the small regularization terms in the objective function of the optimization problem (3.26). By utilizing both the buffer and the storage, the proposed "*hybrid-forwarding distributed*" scheme improves the average throughput by 18.50% compared to "*buffer-forwarding distributed*," and 30.50% compared to "*buffer-forwarding proportional*." "*Storage-forwarding distributed*" obtains the lowest throughput. The comparison of average PSNR is shown in Fig. 3.7(b). The proposed "*hybrid-forwarding distributed*" scheme outperforms "*buffer-forwarding distributed*" by 0.56 dB and "*buffer-forwarding proportional*" by 0.85 dB, respectively.

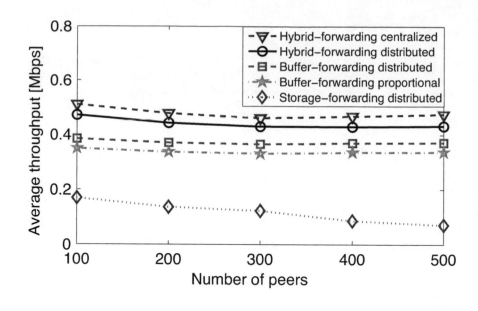

(a)

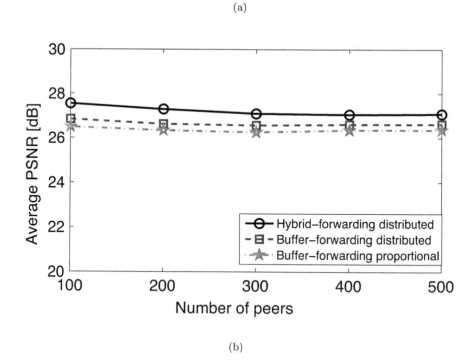

(b)

FIGURE 3.7
Performance comparison with different network sizes: (a) the average throughput, and (b) the average PSNR.

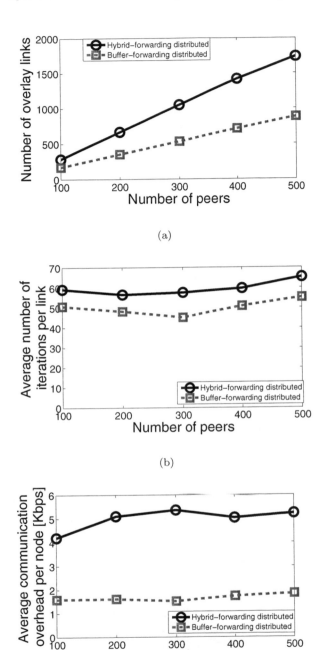

(a)

(b)

(c)

FIGURE 3.8
Comparison of the cost between the hybrid-forwarding architecture and the buffer-forwarding architecture with different network sizes: (a) the number of the overlay links, (b) the average number of iterations per link, and (c) the average communication overhead per node

In the hybrid-forwarding architecture, each peer has not only buffer-forwarding links but also storage-forwarding links. Therefore the number of the links in the hybrid-forwarding architecture is much larger than that in the buffer-forwarding architecture, as shown in Fig. 3.8(a). The proposed algorithm divides the computation load to each link or node. Therefore we can see from Fig. 3.8(b) and Fig. 3.8(c) that the average number of iterations per link and the average communication overhead per node in both architectures do not increase with the network sizes, demonstrating that the proposed algorithm is scalable. Though the communication overhead per node in the hybrid-forwarding architecture is much higher than that in the buffer-forwarding architecture, it only consumes averagely 1.1% of the throughput.

3.3   Streaming Capacity for P2P VoD systems

Most of the P2P VoD systems offer many video channels. Users can choose any of the channels that they are interested in at any time. The P2P VoD systems with multiple channels are called *multi-channel* P2P VoD systems. Depending on the resource correlation, multi-channel P2P VoD systems can be categorized into *independent-channel* P2P VoD systems and *correlated-channel* P2P VoD systems. In an independent-channel P2P VoD system, the peers watching the same channel form an independent overlay, and share the resources with each other exclusively within the overlay. In a correlated-channel P2P VoD system, overlay m formed by the peers watching channel m can be correlated with overlay k formed by the peers watching channel k, such that the peers in overlay m can share the resources not only with other peers in overlay m but also with the peers in overlay k. The resources in a correlated-channel P2P VoD system can be utilized in a better way compared to an independent-channel P2P VoD system.

In P2P VoD systems, users would like to watch the video at a high quality. The streaming rate can be used to indicate the video quality. Let $\mathbf{M}$ denote the set of the channels in a P2P VoD system. If channel $m(m \in \mathbf{M})$ is associated with overlay m, *streaming capacity* for channel m, denoted by c_m, is defined as the maximum streaming rate that can be received by every user in overlay m [32][33]. The *average streaming capacity* c_{avg} for a multi-channel P2P VoD system is defined as: $c_{avg} = \Sigma_{m \in \mathbf{M}} p_m c_m$ where p_m is the priority of channel m, and $\Sigma_{m \in \mathbf{M}} p_m = 1$. The priority of a channel can be set by the service provider. For example, the more expensive channel can be assigned a higher priority.

Streaming capacity in multi-channel P2P VoD systems is a challenging problem. The average streaming capacity in a multi-channel P2P VoD system is dependent on the number of peers, the playback time and the bandwidth of each peer, the server capacity, the overlay construction, and the resource allocation. Optimal resource allocation in a multi-channel P2P VoD system is expected to improve the streaming capacity. However, resource allocation in multi-channel P2P VoD systems is quite challenging due to the following reasons. 1) Peers have heterogeneous upload and download bandwidths and different playback progress. 2) Each channel is heterogeneous in terms of available resources, since the number of the peers in each channel is different. 3) Peers may leave or join a channel dynamically.

Streaming capacity in P2P live systems has been examined in the recent literature [32][33][10][34]. In [32], the streaming capacity problem is formulated into an optimization problem, which maximizes the streaming rate that can be supported by a multi-tree based overlay. Sengupta et al. provide a taxonomy of sixteen problem formulations on streaming capacity, depending on whether there is a single P2P session or there are multiple concurrent sessions, whether the given topology is a full mesh graph or an arbitrary graph, whether the number of peers a node can have is bounded or not, and whether there are non-receiver relay nodes or not [33]. Liu et al. analyze the performance bounds for minimum server load, maximum streaming rate, and minimum tree depth in tree-based P2P live systems, respectively [10]. The streaming capacity under node degree bound is investigated in [34]. Streaming capacity for a single channel in P2P VoD systems has been studied in [35][36][37]. In [35][37], the streaming capacity for a single channel is formulated into an optimization problem which maximizes the streaming rate under the peer bandwidth constraints. The throughput maximization problem in a scalable P2P VoD system is studied in [38]. Helpers have been proposed in P2P systems to improve the system performance [39][40][36]. In P2P VoD systems, each additional helper increases the system upload capacity, thus offloading the server burden [40]. In [36], the algorithms on helper assignment and rate allocation are proposed to improve the streaming capacity for P2P VoD systems.

Cross-channel resource sharing has been recently studied in multi-channel P2P streaming systems [41][42][43][44][45]. In [41], Wu et al. propose an online server capacity provisioning algorithm to adjust the server capacities available to each of the concurrent channels, taking into account the number of peers, the streaming quality, and the priorities of channels. In [42], a View-Upload Decoupling (VUD) scheme is proposed to decouple what a peer uploads from what it views, bringing stability to multi-channel P2P streaming systems and enabling cross-channel resource sharing. In [43], Wu et al. develop infinite-server queueing network models to analytically study the performance of multi-channel P2P live streaming systems. In [44], the bandwidth satisfaction ratio is used to compare three bandwidth allocation schemes, namely the Naive Bandwidth allocation Approach (NBA), the Passive Channel-aware bandwidth allocation Approach (PCA) and the Active Channel-aware bandwidth allocation Approach (ACA), in multi-channel P2P streaming systems. In [45], Zhao et al. investigate the streaming capacity in multi-channel P2P live streaming systems when each peer can only connect to a small number of neighbors.

In this section, we examine the streaming capacity problem for P2P VoD systems. We first investigate the streaming capacity for an independent-channel P2P VoD system, in which we find the streaming capacity for a single channel by optimizing the intra-channel resource allocation in a distributed manner. We then investigate the streaming capacity for a correlated-channel P2P VoD system, in which we find the average streaming capacity for multiple channels by optimizing both intra-channel and cross-channel resource allocation.

3.3.1   Streaming Capacity for an Independent-Channel P2P VoD System

In an independent-channel P2P VoD system, the peers watching the same channel form an overlay. We assume that the server upload bandwidth allocated to channel m, denoted by s_m, is pre-determined. An independent-channel P2P VoD system is illustrated in Fig. 3.9. The peers within the same overlay re-distribute the video content to each other. A peer belonging to overlay m only caches the video content of channel m, and it does not serve any video content to any peer outside overlay m. In other words, each overlay in an independent-channel P2P VoD System is an isolated subsystem. Therefore, the streaming capacity of a channel is independent of that of another channel. The streaming capacity problem in an independent-channel P2P VoD system is to find the streaming capacity of each individual channel, respectively. We will next focus on the streaming capacity for a single channel.

The streaming capacity for a single channel depends on the constructed overlay. In a P2P VoD system, each peer maintains a buffer to cache the recently received packets in order to smoothen the payback and serve other peers. There is a *parent-child relationship* between two peers (e.g., peer k and peer j) if the two peers satisfy the following two conditions: 1) peer k (the parent) has an earlier playback progress than peer j (the child), and 2) peer k is buffering the segment(s) which are being requested by peer j.

A parent-child relationship is *implemented*, if an overlay link has been established from the parent to the child. A *complete overlay* for a channel is defined as the overlay in which all of the available parent-child relationships have been implemented, and an *incomplete overlay* for a channel is defined as the overlay in which only a part of the available parent-child relationships have been implemented [37]. An example of complete overlay is illustrated in Fig. 3.10. As shown in Fig. 3.10, peer 3 is the parent of peer 2, and peer 5 is the parent of peer 4. There is no buffer overlap between peer 3 and peer 4, which means that peer 4 is not a parent of peer 3. The overlay in Fig. 3.10 is a complete overlay since all of the available parent-child relationships have been implemented. Compared to the incomplete overlay, the complete overlay contains more overlay links, thus supporting a higher streaming capacity. Therefore, we will formulate the streaming capacity problem based on the complete overlay. The method for overlay construction is not a study fo-

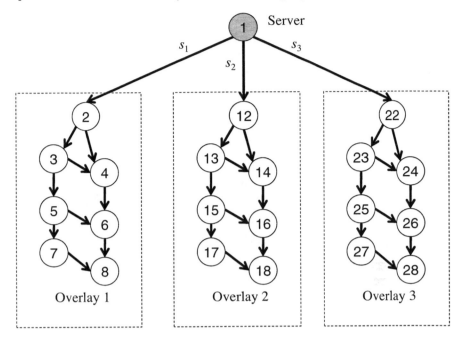

FIGURE 3.9

Illustration of an independent-channel P2P VoD system

cus of this paper. A complete overlay can be constructed using the existing approaches [6].

The set of the overlays (channels) in the independent-channel P2P VoD system is denoted by $\mathbf{M}$. Since each channel is independent in terms of resources, we can just study the streaming capacity of channel m ($\forall m \in \mathbf{M}$).

Overlay m in a P2P VoD system can be modeled as a directed graph $\mathbf{G}^{(m)} = (\mathbf{N}^{(m)}, \mathbf{L}^{(m)})$, where $\mathbf{N}^{(m)}$ is the set of nodes and $\mathbf{L}^{(m)}$ is the set of directed overlay links. Peer 1 is defined as the server. The relationship between a node and its outgoing links is represented with a matrix $\mathbf{A}^{(m)+}$, whose elements are given by

$$a_{il}^{(m)+} = \begin{cases} 1, & \text{if link } l \text{ is an outgoing link from node } i, \\ 0, & \text{otherwise.} \end{cases} \tag{3.27}$$

The relationship between a node and its incoming links is represented with a matrix $\mathbf{A}^{(m)-}$, whose elements are given by

$$a_{il}^{(m)-} = \begin{cases} 1, & \text{if link } l \text{ is an incoming link into node } i, \\ 0, & \text{otherwise.} \end{cases} \tag{3.28}$$

The upload capacity of peer i is denoted by $O_i^{(m)}$. Upload capacity is typically the bottleneck in P2P systems. We consider *upload constraint* at

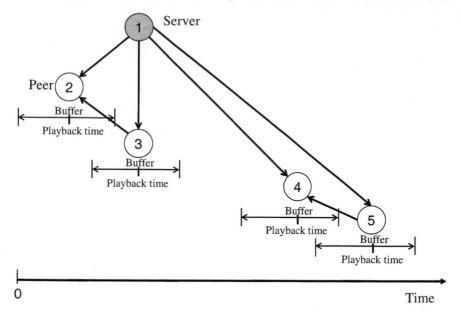

FIGURE 3.10

Illustration of the complete overlay in an independent-channel P2P VoD system

peer $i(\forall i \in \mathbf{N}^{(\mathbf{m})})$, which is given by $\sum_{l \in \mathbf{L}^{(\mathbf{m})}} a_{il}^{(m)+} x_l^{(m)} \leq O_i^{(m)}$ where $x_l^{(m)}$ is the link rate at link l of overlay m. The upload constraint represents that the total outgoing rate from peer i is no larger than its upload capacity $O_i^{(m)}$. The streaming rate for channel m is denoted by $r^{(m)}$. Each peer except the server (e.g., Peer 1) receives the streaming rate $r^{(m)}$, which can be expressed by $\sum_{l \in \mathbf{L}} a_{il}^{(m)-} x_l^{(m)} = f_i^{(m)} r^{(m)}, \forall i \in \mathbf{N}^{(\mathbf{m})}$, where $f_i^{(m)} = 0$ if $i = 1$, or $f_i^{(m)} = 1$ otherwise. The *download constraint* is given by $r^{(m)} \leq \min_{i \in \mathbf{N}^{(\mathbf{m})}} I_i^{(m)}$ where $I_i^{(m)}$ is the download capacity of peer i. From the download constraint, we can see that the maximal streaming rate is limited by the minimal download capacity among the peers. Therefore, the users with a very low download bandwidth should not be admitted into the P2P VoD system in order to maintain a high streaming capacity.

The streaming capacity for channel m is stated as: to maximize the streaming rate $r^{(m)}$ that can be received by every user in channel m by optimizing the streaming rate $r^{(m)}$ and the link rate $x_l^{(m)} (\forall l \in \mathbf{L}^{(\mathbf{m})})$ under the upload constraint at each peer. Mathematically, the problem can be formulated as follows.

$$
\begin{aligned}
\text{maximize} \quad & r^{(m)} \\
\text{subject to} \quad & \sum_{l \in \mathbf{L(m)}} a_{il}^{(m)-} x_l^{(m)} = f_i^{(m)} r^{(m)}, \quad \forall i \in \mathbf{N^{(m)}}, \\
& \sum_{l \in \mathbf{L(m)}} a_{il}^{(m)+} x_l^{(m)} \leq O_i^{(m)}, \qquad \forall i \in \mathbf{N^{(m)}}, \\
& x_l^{(m)} \geq 0, \qquad\qquad\qquad\quad \forall l \in \mathbf{L^{(m)}}, \\
& 0 \leq r^{(m)} \leq \min_{i \in \mathbf{N(m)}} I_i^{(m)}.
\end{aligned}
\tag{3.29}
$$

The optimization problem in (3.29) is a **LP**. It can be solved in a centralized way using the interior point method [21]. However, the centralized solution is not scalable. Therefore we will develop a distributed algorithm using the primal-dual method to solve the streaming capacity problem.

In order to develop a distributed algorithm, we change the problem (3.29) into a proximal optimization problem as follows.

$$
\begin{aligned}
\text{minimize} \quad & -r^{(m)} + \varepsilon (r^{(m)})^2 + \varepsilon \sum_{l \in \mathbf{L(m)}} (x_l^{(m)})^2 \\
\text{subject to} \quad & \sum_{l \in \mathbf{L(m)}} a_{il}^{(m)-} x_l^{(m)} = f_i^{(m)} r^{(m)}, \forall i \in \mathbf{N^{(m)}}, \\
& \sum_{l \in \mathbf{L(m)}} a_{il}^{(m)+} x_l^{(m)} \leq O_i^{(m)}, \forall i \in \mathbf{N^{(m)}}, \\
& x_l^{(m)} \geq 0, \forall l \in \mathbf{L^{(m)}}, \\
& 0 \leq r^{(m)} \leq \min_{i \in \mathbf{N(m)}} I_i^{(m)},
\end{aligned}
\tag{3.30}
$$

where $\varepsilon(\varepsilon > 0)$ is called the *regularization factor*. When ε is small enough, the solution for the problem in (3.30) is arbitrarily close to the solution for the original streaming capacity problem (3.29). The optimization problem (3.30) is a convex optimization problem with a quadratic objective function and the linear constraints [19]. Therefore, we can develop a distributed algorithm to solve the optimization problem (3.30) based on dual decomposition [20].

3.3.2 Streaming Capacity for a Correlated-Channel P2P VoD System

In a correlated-channel P2P VoD system, the resources among different channels can be shared with each other. The average streaming capacity in a correlated-channel P2P VoD system can be obtained by optimizing both the intra-channel resource allocation and the cross-channel resource allocation. The optimization of intra-channel resource allocation for a single channel has been presented in Section 3.3.1. In this section, we will focus on the cross-channel resource allocation. We will first optimize the server upload allocation among channels to maximize the average streaming capacity in Section 3.3.2.1, and then further utilize cross-channel peer upload bandwidth to improve the average streaming capacity in Section 3.3.2.2.

3.3.2.1 Optimization of Server Upload Allocation among Channels

The server upload allocated for channel m is denoted by s_m. In an independent-channel P2P VoD system, the server upload allocation for each

channel is pre-determined. Therefore the server upload bandwidth is not utilized in an optimal way. In this subsection, we treat $\{s_m, \forall m \in \mathbf{M}\}$ as variables, and optimize them to maximize the average streaming capacity for a
correlated-channel P2P VoD system.

The optimization of server upload allocation for a correlated-channel P2P
VoD system is stated as: to maximize the average streaming rate by optimizing the server upload allocation, the link rates, and the streaming rate, for
each channel, subject to the upload constraint at each peer. Since maximizing
the average streaming rate is equivalent to minimizing the negative average
streaming rate, we formulate the problem into a minimization problem as
follows.

$$
\begin{aligned}
\text{minimize} \quad & -\Sigma_{m \in \mathbf{M}} p_m r^{(m)} \\
\text{subject to} \quad & \Sigma_{l \in \mathbf{L^{(m)}}} a_{il}^{(m)-} x_l^{(m)} = f_i^{(m)} r^{(m)}, \\
& \qquad\qquad \forall i \in \mathbf{N^{(m)}}, \forall m \in \mathbf{M}, \\
& \Sigma_{l \in \mathbf{L^{(m)}}} a_{il}^{(m)+} x_l^{(m)} \leq O_i^{(m)}, \\
& \qquad\qquad \forall i \in \mathbf{N^{(m)}}, \forall m \in \mathbf{M}, \\
& x_l^{(m)} \geq 0, \qquad\qquad \forall l \in \mathbf{L^{(m)}}, \\
& \Sigma_{m \in \mathbf{M}} s_m \leq s_T, \\
& s_m \geq 0, \qquad\qquad \forall m \in \mathbf{M}, \\
& 0 \leq r^{(m)} \leq \min_{i \in \mathbf{N^{(m)}}} I_i^{(m)}, \forall m \in \mathbf{M},
\end{aligned}
\tag{3.31}
$$

where p_m is the priority for channel m, and s_T is the server upload capacity.
The optimization problem in (3.31) is an LP. The optimization variables in the
optimization problem (3.31) are the server upload allocation s_m, the link rate
vector $\mathbf{x^{(m)}}$, and the streaming rate $r^{(m)}$, for channel m, $\forall m \in \mathbf{M}$. The number of the optimization variables is increased with the number of the channels
and the number of the links in each overlay. In order to solve the optimization efficiently, we use dual decomposition [20] to decompose the optimization
problem (3.31) into multiple subproblems, each of which is associated with a
channel.

We introduce a dual variable λ for the inequality constraint $\Sigma_{m \in \mathbf{M}} s_m \leq s_T$. The Lagrangian corresponding to the primal problem (3.31) is given by
$L(\mathbf{s}, \mathbf{x}, \mathbf{r}, \lambda) = -\Sigma_{m \in \mathbf{M}} p_m r^{(m)} + \lambda(\Sigma_{m \in \mathbf{M}} s_m - s_T)$. Then the Lagrange dual
function [19] is given by $G(\lambda) = \min L(\mathbf{s}, \mathbf{x}, \mathbf{r}, \lambda) = \Sigma_{m \in \mathbf{M}} \min(-p_m r^{(m)} + \lambda s_m) - \lambda s_T$.

The Lagrange dual problem is to maximize the Lagrange dual function
[19]. That is:

$$
\begin{aligned}
\text{maximize} \quad & G(\lambda) \\
\text{subject to} \quad & \lambda \geq 0.
\end{aligned}
\tag{3.32}
$$

Subgradient method [21] is used to solve the Lagrange dual problem
(3.32). The dual variable λ is updated at the $(k+1)^{th}$ iteration by $\lambda^{(k+1)} = \max\{0, \lambda^{(k)} - \theta^{(k)}(s_T - \Sigma_{m \in \mathbf{M}} s_m^{(k)})$ where $\theta^{(k)}$ is the step size at the kth iteration. In order to guarantee the convergence, the sequence of the step sizes
is required to satisfy the non-summable diminishing rule [20].

At the kth iteration, the primal variables $(s_m, \mathbf{x}^{(m)}, r^{(m)})$ for channel m are obtained by solving the following optimization problem.

$$
\begin{aligned}
\text{minimize} \quad & -p_m r^{(m)} + \lambda s_m \\
\text{subject to} \quad & \sum_{l \in \mathbf{L}^{(m)}} a_{il}^{(m)-} x_l^{(m)} = f_i^{(m)} r^{(m)}, \quad \forall i \in \mathbf{N}^{(m)}, \\
& \sum_{l \in \mathbf{L}^{(m)}} a_{il}^{(m)+} x_l^{(m)} \le O_i^{(m)}, \quad \forall i \in \mathbf{N}^{(m)}, \\
& x_l^{(m)} \ge 0, \quad \forall l \in \mathbf{L}^{(m)}, \\
& 0 \le r^{(m)} \le \min_{i \in \mathbf{N}^{(m)}} I_i^{(m)}, s_m \ge 0.
\end{aligned}
\tag{3.33}
$$

The optimization problem (3.31) is decomposed into $|M|$ subproblems where $|M|$ is the number of the channels. Subproblem m, represented in (3.33), is associated with channel m. Subproblem m, $\forall m \in \mathbf{M}$, can be solved with a distributed algorithm.

3.3.2.2 Cross-Channel Sharing of Peer Upload Bandwidth

Though the server upload allocation among channels is optimized, the cross-channel resources have not yet been fully utilized. In each channel, there are a number of under-utilized peers, which can be utilized by the other channels. For example, the leaf nodes of overlay m contribute zero upload bandwidth to the channel because they have no outgoing links. We can utilize the upload bandwidth of the leaf nodes in overlay m to serve the peers in another channel (e.g., channel k), thus improving the streaming capacity of channel k. However, it is challenging to establish the cross-channel links to enable the cross-channel sharing of peer upload bandwidth.

In this book, we propose a scheme for cross-channel peer upload sharing. We introduce the concept of *cross-channel helpers*. A *cross-channel helper* is the peer who uses its remaining upload bandwidth to help other peers in another channel. Cross-channel helpers are chosen from the peers who have a remaining upload bandwidth greater than a threshold.

Suppose that peer i in channel m is a cross-channel helper serving segment j of channel k, and it has a remaining upload bandwidth b_i. We denote the set of peers watching segment j in channel k by $\mathbf{H}_j^{(k)}$. Peer i can download segment j of channel k from the server or the peers who are buffering segment j, at a rate R_i^{IN}, and then output it to peer $h, \forall h \in \mathbf{H}_j^{(k)}$, at a rate R_i^h, respectively. The *bandwidth gain* for peer i is defined as $g_i = (\sum_{h \in \mathbf{H}_j^{(k)}} R_i^h) / R_i^{IN}$. In order to maximize the bandwidth gain g_i under the flow constraint $R_i^h \le R_i^{IN}, \forall h \in \mathbf{H}_j^{(k)}$, and the bandwidth constraint $\sum_{h \in \mathbf{H}_j^{(k)}} R_i^h \le b_i$, the optimal download rate at peer i is $R_i^{IN*} = b_i / |\mathbf{H}_j^{(k)}|$ where $|\mathbf{H}_j^{(k)}|$ is the number of the peers watching segment j of channel k, and the optimal outgoing rate to peer $h, \forall h \in \mathbf{H}_j^{(k)}$, is $R_i^{h*} = R_i^{IN*}$. The bandwidth gain should be larger than 1, otherwise the cross-channel helper consumes a larger bandwidth than it contributes. Fig. 3.11 illustrates a correlated-channel P2P VoD system, in which peer 8 is a cross-channel helper, who downloads a segment from the server and then forwards it to peers 15, 16, and 17, respectively.

The proposed scheme for cross-channel peer upload sharing is described as follows.

1. Channel $m(\forall m \in \mathbf{M})$ finds a partner, channel $k(k \neq m, k \in \mathbf{M})$, to help each other. We propose a *resource-balance scheme* to find the channel pair, described as follows. 1) Calculate the amount of the total remaining bandwidth B_m^{re} for channel $m(\forall m \in \mathbf{M})$, the average amount of the total remaining bandwidth is given by $B_{avg}^{re} = (\sum_{m \in \mathbf{M}} B_m^{re})/|\mathbf{M}|$ where $|\mathbf{M}|$ represents the number of channels. 2) Given that the set of unmatched channels is $\mathbf{M}^{un}$, the partner of channel m is determined by $k = \underset{j \in \mathbf{M}^{un}}{\arg \min}(B_j^{re} - B_{avg}^{re})^2$.

2. Determine the set of cross-channel helpers in channel m, denoted by $\mathbf{Y}^{(m)}$, by choosing the peers who have a remaining upload bandwidth greater than a threshold b_{th}. Sort $\mathbf{Y}^{(m)}$ in a descending order based on the remaining upload bandwidth.

3. Determine the *demanding segment set* in channel k, denoted by $\mathbf{W}^{(k)}$, by choosing $|\mathbf{Y}^{(m)}|$ segments which are watched by the largest number of peers. Sort $\mathbf{W}^{(k)}$ in a descending order based on the number of watching peers.

4. Assign the ith cross-channel helper in $\mathbf{Y}^{(m)}$ to serve the peers watching the ith segment in $\mathbf{W}^{(k)}$. Determine the incoming rate and outgoing rates at the ith cross-channel helper to maximize the bandwidth gain.

5. After allocating the rate for each of the cross-channel links as in Step 4), re-optimize the server upload allocation and the link rates within each overlay by solving the optimization problem (3.31).

Peer dynamics have an impact on the streaming capacity in correlated-channel P2P VoD systems. First, the peers may leave or join a channel dynamically. Second, the cross-channel helpers may leave the channel, which causes the disconnection of the cross-channel links. To handle the dynamic conditions, the optimizations of intra-channel resource allocation and cross-channel resource allocation need to be performed in a discrete-time manner, in which the peers and the overlay are assumed to remain unchanged during a time slot, and the algorithms for resource allocation are performed at the beginning of each time slot.

3.3.3 Simulation Results

In the simulations, we use two classes of peers: cable/DSL peers and Ethernet peers. Cable/DSL peers take 85% of the total peer population with download

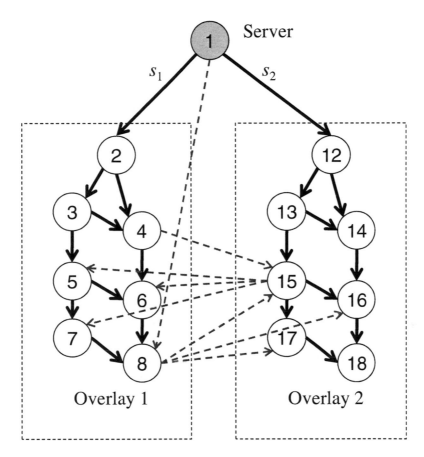

FIGURE 3.11
Illustration of a correlated-channel P2P VoD system

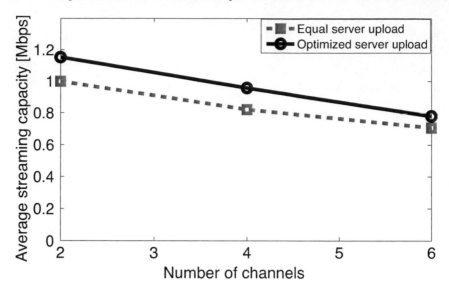

FIGURE 3.12
Comparison of average streaming capacity with different number of channels

capacity uniformly distributed between 0.9 Mbps and 1.5 Mbps and upload capacity uniformly distributed between 0.3 Mbps and 0.6 Mbps. Ethernet peers take the remaining 15% of the total peer population with both upload and download capacities uniformly distributed between 1.5 Mbps and 3.0 Mbps. The length of the video is 60 minutes, which is evenly divided into 60 segments. Each peer maintains a buffer with a capacity of 5 segments. The playback time of each peer is randomly distributed between 0 and 60 minutes. The priorities for all channels are equal.

Fig. 3.12 compares the average streaming capacity between the *equal server upload* scheme, in which each channel is allocated an equal server upload, and the *optimized server upload* scheme, in which the server upload for each channel is optimized as described in Section 3.3.2.1. The number of the peers in a channel is uniformly distributed between 10 and 150. The server upload capacity is 45.0 Mbps. The optimized server upload scheme improves the average streaming capacity by 14.1% in average compared to the equal server upload scheme.

We show in Fig. 3.13 the impact of server upload capacity to the average streaming capacity for a P2P VoD system with 2 channels. The first channel has 40 peers, and the second one has 120 peers. As shown in Fig. 3.13, the average streaming capacity is increased with the server upload capacity. The improvement brought by the optimized server upload scheme is larger when the server upload capacity is larger.

Joint consideration of server upload optimization and cross-channel peer

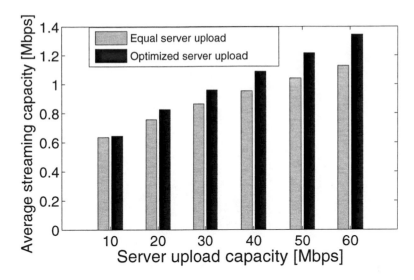

FIGURE 3.13
Impact of server upload capacity to the average streaming capacity

upload sharing can improve the performance in a correlated P2P VoD system. There are two channels with total 200 peers in the P2P VoD system. The server upload capacity is 20.0 Mbps. We vary the number of the peers in channel 1 from 20 to 140, and evaluate two metrics: peer upload utilization ratio and average streaming capacity. The *peer upload utilization ratio* is defined by $\beta = \Sigma_{i \in \mathbf{N}, i \neq 1} v_i / \Sigma_{i \in \mathbf{N}, i \neq 1} O_i$ where $\Sigma_{i \in \mathbf{N}, i \neq 1} v_i$ represents the sum of the outgoing rates from all peers except the server, and $\Sigma_{i \in \mathbf{N}, i \neq 1} O_i$ represents the sum of the upload capacities of all peers except the server. Due to a better utilization of cross-channel resources, joint consideration of server upload optimization and cross-channel peer upload sharing improves the peer upload utilization ratio by 7.7% on average, as shown in Fig. 3.14(a), and improves the average streaming capacity by 0.04 Mbps on average, as shown in Fig. 3.14(b), compared to the *optimized server upload scheme*.

3.4 Summary

In this chapter, we have presented the overview of P2P streaming systems. In tree-based P2P live streaming systems, a single application-layer tree or multiple application-layer trees are constructed to deliver the video streams. In mesh-based P2P live streaming systems, each peer finds its neighbors and then exchanges data with them. Different from live streaming services in which

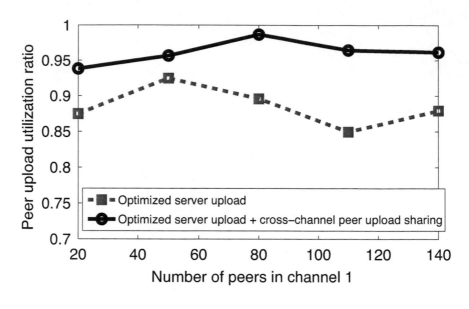

(a)

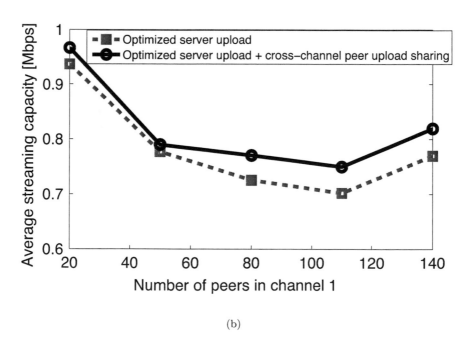

(b)

FIGURE 3.14
Performance improvement brought by cross-channel peer upload sharing: (a) peer upload utilization ratio, and (b) average streaming capacity

each user watches almost the same segment of the video, VoD services allow users to watch any point of video at any time.

We have examined two resource allocation problems, which are: 1) distributed throughput maximization for scalable P2P VoD systems, and 2) streaming capacity for P2P VoD systems. In the throughput maximization problem, we have presented distributed algorithms to maximize the throughput for buffer-forwarding P2P VoD systems and hybrid-forwarding P2P VoD systems, respectively. In the streaming capacity problem, we have optimized both the intra-channel resource allocation and cross-channel resource allocation to improve the streaming capacity for P2P VoD systems. We have demonstrated in the simulations that the system performance can be improved by better utilization of the resources of the peers.

Bibliography

[1] http://www.youtube.com

[2] Gomes L. (2006) Will all of us get our 15 minutes on a YouTube video? In: *Wall Street Journal*

[3] Hefeeda M., Bhargava B., Yau D. (2004) A Hybrid Architecture for Cost-Effective On-Demand Media Streaming, In: *Elsevier Computer Networks*, vol. 44, no. 3, pp. 353–382

[4] http://www.akamai.com/

[5] X. Zhang, J. Liu, B. Li, and T. P. Yum, "CoolStreaming/DONet: A Data-Driven Overlay Network for Efficient Live Media Streaming," in *Proc. of IEEE INFOCOM*, vol. 3, pp. 2102–2111, Mar. 2005.

[6] H. Chi, Q. Zhang, J. Jia, and X. Shen, "Efficient search and scheduling in P2P-based media-on-demand streaming service," *IEEE Journal on Selected Areas in Communications*, vol. 3, pp. 1467–1472, Jun. 2006.

[7] I. Lee and L. Guan, "Centralized peer-to-peer streaming with layered video," in *Proc. of IEEE ICME*, vol. 1, pp. 513–516, Jul. 2003.

[8] Chu Y, Rao S, Seshan S, Zhang H (2002) A Case for End System Multicast, In: IEEE Journal on Selected Areas in Communications, vol. 20, no. 8, pp. 1456–1471

[9] Castro M, Druschel P, Kermarrec A-M, Nandi A, Rowstron (2003) SplitStream: high-bandwidth multicast in cooperative environments, In: Proc. of ACM SOSP, pp. 298–313

[10] Liu Y, Guo Y, Liang C (2008) A survey on peer-to-peer video streaming systems, In: Springer Journal on Peer-to-Peer Networking and Applications, no. 1, pp. 18–28

[11] Zhang M, Luo J G, Zhao L, Yang S Q (2005) A peer-to-peer network for live media streaming using a push-pull approach, In: Proc. of ACM MM, pp. 287–290

[12] Z. Li and A. Mahanti, "A Progressive Flow Auction Approach for Low-Cost On-Demand P2P Media Streaming," in *Proc. of ACM QShine*, Aug. 2006.

[13] C. Huang, J. Li, and K. W. Ross, "Peer-Assisted VoD: Making Internet Video Distribution Cheap," in *Proc. of IPTPS*, Feb. 2007.

[14] Yiu W P, Jin X, Chan S H (2007) VMesh: Distributed segment storage for peer-to-peer interactive video streaming, In: IEEE Journal on Selected Areas in Communications, vol. 25, no. 9, pp. 1717–1731

[15] X. Xu, Y. Wang, S. P. Panwar, and K. W. Ross, "A Peer-to-Peer Video-on-Demand System using Multiple Description Coding and Server Diversity," in *Proc. of IEEE ICIP*, vol. 3, pp. 1759–1762, Oct. 2004.

[16] He Y, Lee I, Ling G (2008) Distributed throughput maximization in hybrid-forwarding P2P VoD applications, In: Proc. of IEEE ICASSP, pp. 2165–2168

[17] Y. Shen, Z. Liu, S. S. Panwar, K. W. Ross, and Y. Wang, "Streaming Layered Encoded Video Using Peers," in *Proc. of ICME*, Jul. 2005.

[18] J. Byers, J. Considine, M. Mitzenmacher, and S. Rost, "Informed Content Delivery Across Adaptive Overlay Networks," *IEEE/ACM Transactions on Networking*, vol. 12, no. 5, pp. 767–780, Oct. 2004.

[19] C. Wu and B. Li, "rStream: Resilient Peer-to-Peer Streaming with Rate-less Codes," in *Proc. of ACM MM*, pp. 307–310, Nov. 2005.

[20] R. Ahlswede, N. Cai, S.-Y. R. Li and R. W. Yeung, "Network information flow," *IEEE Transactions on Information Theory*, vol. 46, pp. 1204–1216, Jul. 2000.

[21] J. Liang and K. Nahrstedt, "DagStream: Locality Aware and Failure Resilient Peer-to-Peer Streaming," in *Proc. of SPIE MMCN*, vol. 6071, pp. 60710L -1-60710L-15, Jan. 2006.

[22] M. Hefeeda, A. Habib, B. Botev, D. Xu, and B. Bhargava, "PROMISE: Peer-to-Peer Media Streaming Using CollectCast," in *Proc. of ACM MM*, pp. 45–54, Nov. 2003.

[23] R. J. Vanderbei, *Linear Programming: Foundations and Extensions,* 2nd Edition, Springer Press, 2001.

[24] S. Boyd and L. Vandenberghe, *Convex Optimization,* Cambridge University Press, 2004.

[25] D. Palomar and M. Chiang, "A tutorial on decomposition methods and distributed network resource allocation," *IEEE Journal on Selected Areas in Communications,* vol. 24, no. 8, pp. 1439–1451, Aug. 2006.

[26] D. P. Bertsekas, A. Nedic, and A. E. Ozdaglar, *Convex Analysis and Optimization,* Athena Scientific, 2003.

[27] H. Schwarz, D. Marpe, and T. Wiegand, "SNR-scalable extension of H.264/AVC," in *Proc. of IEEE ICIP,* vol. 5, pp. 3113–3116, Oct. 2004.

[28] B. Chen, X. Liu, Z. Zheng, and H. Jin, "A measurement study of a peer-to-peer video-on-demand system," in *Proc. of IPTPS,* Feb. 2007.

[29] T. T. Do, K. A. Hua, and M. A. Tantaoui, "P2VoD: Providing Fault Tolerant Video-on-Demand Streaming in Peer-to-Peer Environment," in *Proc. of IEEE ICC,* vol. 25, no. 1, pp. 119–130, Jan. 2004.

[30] Y. Cui, B. Li, and K. Nahrstedt, "oStream: asynchronous streaming multicast," *IEEE Journal on Selected Areas in Communications,* vol. 22, no. 1, pp. 91–106, Jan. 2004.

[31] W. P. Yiu, X. Jin and S. H. Chan, "Distributed storage to support user interactivity in peer-to-peer video streaming," in *Proc. of IEEE ICC,* vol. 1, pp. 55–60, Jun. 2006.

[32] S. Sengupta, S. Liu, M. Chen, M. Chiang, J. Li, and P. A. Chou, "Streaming Capacity in Peer-to-Peer Networks with Topology Constraints," *Microsoft Research Technical Report,* 2008.

[33] S. Sengupta, S. Liu, M. Chen, M. Chiang, J. Li, and P.A. Chou, "Peer-to-Peer Streaming Capacity," *IEEE Transactions on Information Theory,* vol. 57, no. 8, pp. 5072–5087, Aug. 2011.

[34] S. Liu, M. Chen, S. Sengupta, M. Chiang, J. Li, and P.A. Chou, "P2P Streaming Capacity under Node Degree Bound," in *Proc. of IEEE ICDCS,* pp. 587–598, 2011.

[35] Y. He and L. Guan, "Streaming capacity in P2P VoD systems," in *Proc. of IEEE ISCAS,* pp. 742–745, May 2009.

[36] Y. He and L. Guan, "Improving the streaming capacity in P2P VoD systems with helpers," in *Proc. of IEEE ICME,* pp. 790–793, Jul. 2009.

[37] Y. He and L. Guan, "Solving streaming capacity problems in P2P VoD systems," *IEEE Transactions on Circuits and Systems for Video Technology,* vol. 20, no. 11, pp. 1638–642, Nov. 2010.

[38] Y. He, I. Lee, and L. Guan, "Distributed throughput maximization in P2P VoD applications," *IEEE Transactions on Multimedia,* vol. 11, no. 3, pp. 509–522, Apr. 2009.

[39] J. Wang, C. Yeo, V. Prabhakaran, and K. Ramchandran, "On the Role of Helpers in Peer-to-Peer File Download Systems: Design, Analysis and Simulation," in *Proc. of IPTPS,* Feb. 2007.

[40] P. Garbacki, D. Epema, J. Pouwelse, M. van Steen, "Offloading Servers with Collaborative Video on Demand," in *Proc. of IPTPS,* Feb. 2008.

[41] C. Wu, B. Li, and S. Zhao, "Multi-channel Live P2P Streaming: Refocusing on Servers," in *Proc. of IEEE INFOCOM,* pp. 1355–1363, Apr. 2008.

[42] D. Wu, C. Liang, Y. Liu and K. W. Ross, "View-upload decoupling: a redesign of multi-channel P2P video systems," in *Proc. of IEEE INFOCOM,* pp. 2726–2730, Apr. 2009.

[43] D. Wu, Y. Liu, and K.W. Ross, "Queuing Network Models for Multi-Channel P2P Live Streaming Systems," in *Proc. of IEEE INFOCOM,* pp. 73–81, 2009.

[44] M. Wang, L. Xu, and B. Ramamurthy, "Linear Programming Models For Multi-Channel P2P Streaming Systems," in *Proc. of IEEE INFOCOM,* 2010.

[45] C. Zhao, X. Lin, and C Wu, "The streaming capacity of sparsely-connected P2P systems with distributed control," in *Proc. of IEEE INFOCOM,* pp. 1449–1457, 2011.

[46] Y. He and L. Guan, "Streaming capacity in multi-channel P2P VoD systems," in *Proc. of IEEE ISCAS,* pp. 1819–1822, May 2010.

4

Prefetching Scheme and Substream Allocation in P2P VoD Applications

CONTENTS

Buffer and storage are two major resources in the Peer-to-Peer (P2P) Video-on-Demand (VoD) applications. The peers replicate the video segments in their storage or cache video segments in their buffer, such that these segments can be served to other peers. If the resources (e.g., buffer and storage) are utilized in an optimal way, we can greatly improve the P2P VoD performances.

In this chapter, we study the optimal utilization of buffer or storage in P2P VoD applications.

In the first part of this chapter, we present an optimal prefetching framework in P2P VoD applications with guided seeks. In P2P VoD applications, users seek frequently to the positions they are interested in. In this work, we propose the concept of guided seeks. With the guidance, users can perform efficient seeks to the desired positions. The guidance is obtained by collecting the seeking statistics in the previous sessions. It is quite challenging to aggregate the statistics in the distributed P2P networks. We design the hybrid sketches to represent the seeking statistics, thus greatly reducing the time complexity and space complexity. From the collected seeking statistics, we estimate the segment access probability, based on which we develop an optimal prefetching scheme and an optimal cache replacement policy to minimize the expected seeking delay at every viewing position.

In a layered P2P streaming system, each video is encoded into multiple layers. The bit stream in a layer is called a substream, which can be independently replicated in the storage of the peers. It is quite challenging how many copies for each substream should be generated in order to obtain a better performance since each of the layers has a different priority. In the second part of this chapter, we present an optimal substream allocation scheme to tackle this problem. The simulation results show that the proposed allocation scheme enables the system to achieve an overall higher quality compared to the general allocation schemes with fixed allocation percentages.

4.1   Optimal Prefetching Scheme in P2P VoD Applications with Guided Seeks

4.1.1   Introduction

Most of the existing work on P2P-based VoD systems [1, 2] has made an implicit assumption that a user who has joined a streaming session will keep on watching till it leaves or the session fails. This assumption excludes flexible VCR functionalities from VoD systems to make the system design simple. Unfortunately, previous analysis on a large amount of real user viewing logs has shown that users usually do not play the video successively and passively [3]. Instead, users perform seeks quite frequently [3]. The reasons for these seeking behaviors are: 1) some users may feel that the current segment is boring, so they skip away from it; 2) some users may not have sufficient time to watch the whole video and just want to browse some exciting or interesting segments.

In the P2P VoD applications, the frequent seeks pose a great challenge on the playback continuity, which can be measured with the seeking delay, defined as the interval between the time when a segment is requested and the time when the segment is ready for playback. Ideally, we would like to have a zero seeking delay, such that we can view any segment in the video anytime without any interruption. To achieve a zero seeking delay, each segment has to be fetched prior to its playback time. A prefetching scheme prefetches one or multiple segments while the current segment is being played. It turns out to increase the playback continuity [4]. However, most of the existing systems assume that users watch the video sequentially without any seek. Therefore, they use sequential prefetching schemes [4, 5], in which each peer prefetches the segments next to the current segment in sequential order. In the VoD applications with random seeks, the next position the peer will access may not be the segment next to the current segment; it will be probably any segment in the whole video. Therefore, the sequential prefetching scheme does not work well if the peers perform seeks frequently. The prefetching scheme in P2P VoD systems with frequent seeks needs to predict the access probability and then prefetch the appropriate segments.

Random seek is one of the important characteristics in VoD applications. There is much work on the overlay construction with resilience to random seeks. The overlay can be constructed in tree structures, such as P2VoD [7] and oStream [8], or mesh structures [9][10]. Most of them organize the overlay based on the play progress [9][10], such that the peers can forward the recently played content to the followers. Other P2P VoD architectures place the segments of the video on peer storage [11][12]. When a peer jumps to a new position, it can locate the owner of the new segment by looking up the Distributed Hash Table (DHT) [11]. Ring-assisted overlay topology, called RINDY, has been proposed in [13]. In RINDY, a peer can implement fast relocation for random seeks by maintaining some near neighbors and remote neighbors in a set of concentric rings.

The file popularity has been used to design the strategy of file distribution in P2P networks. In Push-to-Peer VoD [14], the video file is proactively pushed to peers to increase content availability and improve the use of peer uplink bandwidth. In [15], an optimization methodology is introduced to replicate and replace files as a function of their popularity to maximize file availability in P2P content distribution. In VMesh [11], a popularity-based segment storage scheme is proposed for P2P VoD applications.

Prefetching is an effective approach to reduce the waiting time for an object in the applications of file transfers and media streaming. In [16], the authors optimize the speculative prefetching scheme by solving a stretched knapsack problem. In [17], a prefetching algorithm with low computational complexity is proposed to minimize the expected waiting time of document retrieval. In P2P streaming applications, Sharma et al. [4] have proposed a distributed prefetching protocol where peers prefetch and store portions of the streaming media ahead of their playback time. Shen et al. [5] have examined a prefetching scheme in P2P VoD applications such that the client peer can tap the reservoir of the prefetched bits while searching for a replacement serving peer. However, these prefetching schemes assume that the peers do the sequential playback without any seek.

The concept of *guided seeks* is presented in this section. A guided seek is different from a random seek in that it is performed based on the segment access information, which is learned from the seeking statistics in the previous and/or concurrent sessions. Typically, the popular segments are visited more frequently, thus the access count of a segment approximately reflects its popularity. With guidance, users can jump to the desired positions in the video quickly and efficiently. For example, a user watching an on-demand soccer match may want to just watch the segments with goals. However, he or she does not know where the scenes with goals are. In the VoD applications with guidance, the system can provide an indicating bar of the segment popularity in the media player, with a grey scale indicating the degree of the popularity for each segment. This user can just drag the progress bar to the position in dark color since the segment in dark color is popular, thus containing the goal scenes with a higher probability. In addition to segment popularity, other information, such as user tags, user ratings, and interest groups, can be collected to guide users in performing efficient seeks.

In order to provide guidance to users, we need to obtain the segment access probability. The segment access probability is a two-dimensional Probability Density Function (PDF), denoted as $P(x, y)$, where x is the start segment of a seek and y is the destination segment of a seek. $P(x, y)$ can be learned from the seeking statistics in the previous and/or concurrent sessions. In a centralized VoD system, a server can be used to collect the seeking statistics. However, it overloads the burden of the server and introduces a single point of failure. Nevertheless, collecting the seeking statistics in distributed P2P networks is a nontrivial task. There are two major challenges: 1) the message size grows linearly with the number of the collected seeking statistics; 2) the duplicate statistics cause an inaccuracy in the estimation of the segment access probability, hence we should avoid the double-counting problem in the statistics collection.

In this section, we present an optimal prefetching framework in P2P VoD applications with guided seeks. The proposed framework is shown in Figure 4.1. It consists of two modules: the seeking statistics aggregation and the optimal prefetching scheme. In the seeking statistics aggregation, we represent the seeking statistics with the *hybrid sketches* to prevent the message size from growing linearly. The peer (e.g., peer i) first retrieves a set of random neighbors (e.g., peers m, n and k) from the tracker, and then gossips the hybrid sketches with them. Based on the aggregated seeking statistics, the peer estimates the segment access probability $P(x, y)$, which is used for the design of the optimal prefetching scheme, and is also used to guide the user to perform efficient seeks. The module of the optimal prefetching scheme takes the segment access probability $P(x, y)$ as input, and then determines the optimal segments for prefetching and the optimal cache replacement policy.

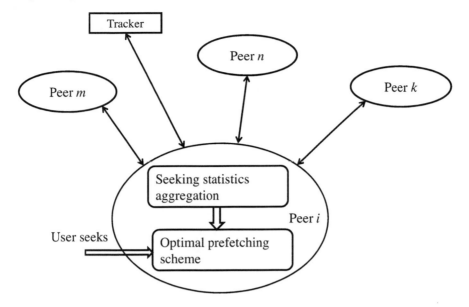

FIGURE 4.1
The proposed framework for P2P VoD applications with guided seeks

4.1.2   Seeking Statistics Aggregation

We describe the seeking statistics aggregation in this section and the optimal prefetching scheme in the next section, respectively. The notations used in this section are summarized in Table 4.1.

In the seeking statistics aggregation, we use gossip protocol to aggregate the information because 1) gossip protocol is robust for peer dynamics, and 2) many P2P live streaming [18] or P2P VoD systems [13] employ gossip protocol to exchange media content. In our framework, gossip occurs periodically. In each round of random gossip, every peer talks to one or more randomly selected neighbors and exchanges the information with them. It turns out that, after approximately $logN$ rounds of computation where N is the number of the peers, all the peers can obtain the global information with a high probability [19].

A video is uniformly divided into M segments; each segment has a length of T. We assume a segment is the smallest unit in the time scale. A peer can either play a segment or skip it. A *seeking behavior* $s(x, y)$ is characterized by the start segment, denoted by x, and the destination segment, denoted by $y (y \neq x)$. The sequential playback is a special seeking behavior, in which $y = x + 1$. A seeking behavior $s(x, y)$ contains two variables. To reduce the number of the variables, we map a seeking behavior $s(x, y)$ into a corresponding *seeking*

Symbol	Definition
N	The number of the peers
M	The number of the segments in the video
$s(x, y)$	A seeking behavior from segment x to segment y
e_v	The seeking record of seeking type v
u_j	The jth seek ID
N_i^T	The number of the seeking types in the data stream at peer i
S_{R_i}	The space to store the data stream at peer i
b	The number of the bits used to represent a seek ID
FM_W	The Flajolet-Martin (FM) sketch of a multi-set W
L	The length of a bitmap
c	The position of the least-significant 0-bit in the FM sketch
N_b	The number of the bitmaps in a FM sketch
$f(x, y)$	The access frequency of seeking behavior $s(x, y)$
$P(x, y)$	The access probability from segment x to segment y
$P(y|\alpha)$	The access probability to segment y at current segment α
r_s	the constant bit rate of the video
T	The viewing time of a segment
s_{seg}	The size of a segment
Y	The whole set containing all the segments in the video
B_i	The cache set at segment i
U_i	The unavailable set at segment i
F_i	The prefetched set at segment i
A_i	The complete set at segment i
G_i	The ongoing set at segment i
t_i	The playback time of segment i
θ_i	The prefetching shift of segment i
τ_j	The seeking delay of segment j
o_i	The upload capacity of segment i
b_d	The download capacity of the receiving peer
r_m	The minimum rate to completely download a segment during the prefetching interval
r_k	The prefetching rate of segment k in the unavailable set
$q_{k|i}$	The Access-probability to Upload-capacity Ratio (AUR) of segment k when segment i is watched
ξ	The maximum number of the segments in the cache set
β	The sampling ratio
λ_β	The compression ratio at sampling ratio β
e_m^β	The counting error of seeking type m at sampling ratio β
ε_β	The Mean Absolute Error (MAE) at sampling ratio β

TABLE 4.1

Notations in the optimal prefetching framework for P2P VoD applications [6] *type* $v \in \{1, 2, \cdots, M(M-1)\}$, which is given by

$$v = \begin{cases} (M-1)(x-1) + y, & \text{if } y < x, \\ (M-1)(x-1) + y - 1, & \text{if } y > x. \end{cases} \qquad (4.1)$$

A peer collects the seeking statistics in the P2P network and puts them into a *data stream*. The data stream at peer i is denoted by R_i, which contains a series of *seeking records* $R_i = \{e_v\}$. The seeking record corresponding to seeking type v is denoted by a set $e_v = \{u_j\}$, where u_j is a unique *seek ID* which can be formed as the concatenation of the peer ID and the sequence number generated by this peer. For example, peer 4 maintains a data stream as $R_4 = \{e_1 = \{2, 4\}, e_3 = \{3, 6, 12, 16, 21, 32, 45, 51, 62\}\}$, in which seeking record e_1 records 2 seeks (seek IDs = 2 and 4) associated with seeking type 1, and seeking record e_3 records 9 seeks corresponding to seeking type 3. In a gossip-based information aggregation, there are risks that a peer may receive duplicate information. By carrying the seek IDs in each seeking record, we can avoid the double-counting problem.

4.1.2.1   Intuitive Approach

We first examine an intuitive approach, in which each peer exchanges each of the seeking records in its data stream with its neighbors via gossips. In the intuitive approach, the procedure of information exchange between seeking record e_v^i at peer i and seeking record e_v^j at peer j is as follows. First, peer i sorts its seek IDs in seeking record e_v^i. Second, peer j searches each seek ID in its seeking record e_v^j, and puts the seek ID into e_v^i if it has not appeared in e_v^i. The intuitive approach can accurately aggregate the information. However, the space complexity and time complexity is large. Let S_i^a denote the average number of seek IDs for a seeking type at peer i, N_i^T denote the number of the seeking types in the data stream at peer i. For the information aggregation between two peers (e.g., peer i and peer j), the time complexity at peer i is $O(N_j^T (S_i^a \log S_i^a + S_j^a \log S_i^a))$. The space to store the data stream R_i at peer i is $S_{R_i} = b \sum_{e_v^i \in R_i} |e_v^i|$, where $|e_v^i|$ denotes the number of seek IDs in seeking record e_v^i, and b is the number of the bits used to represent a seek ID. We can see that the space requirement is linearly scaling with the number of the seek IDs.

4.1.2.2   The Proposed Hybrid Sketches

To reduce the space and time complexity, we employ sketches to represent the data stream at each peer. A sketch is a compact representation of a data stream. It prevents the space from scaling linearly.

The Flajolet-Martin (FM) sketch was introduced for estimating the number of distinct objects in a multi-set in one pass while using only a small amount of space [20]. Given a multi-set W, the FM sketch FM_W of W is a bitmap of L bits. The bitmap length L is given by $L \approx \log_2 n$, where n is the number of distinct elements in the multi-set. The bit of FM_W is denoted by

$FM_W(l)$, $l = 1, \cdots, L$. All the bits in FM_W are initialized to zero. A uniform hash function $h(x)$ maps each element $\omega \in W$ to $h(\omega) \in \{0, \cdots, 2^L - 1\}$. Let $I(\omega)$ denote the position of the least-significant 1-bit in $h(\omega)$. For each element $\omega \in W$, we set $FM_W(I(\omega)) = 1$. If $h(\omega) = 0$, we allocate $I(\omega) = L$. For example, if $L = 5$, $h(\omega) = 6 = 00110$ (in binary), then $I(\omega) = 2$. If ω is the first element inserted into FM_W, the FM bitmap will be $FM_W = 00010$. Since duplicate elements just set the same bit regardless of the order of the insertion, FM sketch is Order-and-Duplicate-Insensitive (ODI).

To estimate the number of distinct elements from an FM sketch, the following approach is used. FM finds the first zero bit of the sketch from the least-significant bit. Let the position of this bit be c. Then the number of distinct elements is estimated as $n = 1.2928 \times 2^c$ [20]. To reduce the approximation error, Flajolet and Martin proposed a technique called Probabilistic Counting with Stochastic Averaging (PCSA) [20]. PCSA applies a second hash function to choose one of the N_b bitmaps and performs the element insertion only to this bitmap. As a result, each bitmap is responsible for approximately n/N_b distinct elements. The number of distinct elements estimated with PCSA is given by $n = 1.2928 N_b 2^{\frac{1}{N_b} \sum_{i=1}^{N_b} c_i}$, where c_i is the position of the first zero bit in bitmap i [20].

PCSA FM sketch is efficient in wrapping a multi-set of large number of elements into a matrix of $N_b L$ bits. However, it is inefficient in representing a set of small elements in terms of space consumption. Moreover, the relative error is quite large if a FM sketch is used to represent a small number of elements. Therefore, we design a *hybrid sketch* to represent the seek IDs in each seeking record. The hybrid sketch is constructed as follows. If the number of seek IDs in a seeking record e_v is less than (or equal to) a threshold γ_{ID}, this seeking record is called a light seeking record, we will just enumerate all seek IDs into the hybrid sketch. On the other hand, the seeking record with the number of seek IDs larger than γ_{ID} is called a heavy seeking record, the corresponding hybrid sketch is a FM sketch, wrapping all the seek IDs in the heavy seeking record into a matrix of $N_b L$ bits.

Each peer (e.g., peer i) maintains a set of hybrid sketches. Peer i periodically updates its hybrid sketches as follows. First, peer i inserts the data stream generated by itself since last update time into hybrid sketches. Second, peer i asks the tracker to get K random neighbors, and then performs pairwise gossip with each of the K neighbors respectively. In the pair-wise gossip with peer j, peer i gets the hybrid sketches from peer j, and updates each of its hybrid sketches. If the hybrid sketch is the enumeration of seek IDs in the light seeking record, the update of the sketch is performed using the intuitive approach. If the hybrid sketch is the FM sketch of the heavy seeking record, the update is performed via bit-wise OR operation.

Let N_i^T denote the number of the seeking records at peer i. Among N_i^T seeking records, the number of the light seeking records is N_i^{Tl}, and the number of the heavy seeking records is N_i^{Th}. Let S_i^{al} denote the average number of seek IDs in a light seeking record, and S_i^{ah} denote the average number of seek Ds in a heavy seeking record. If peer i exchanges hybrid sketches with peer j, the time complexity at peer i is $O(N_j^{Tl}(S_i^{al} \log S_i^{al} + S_j^{al} \log S_i^{al}) + N_j^{Th} N_b)$. If the intuitive approach is used, the time complexity at peer i is $O(N_j^{Tl}(S_i^{al} \log S_j^{al} + S_j^{al} \log S_i^{al}) + N_j^{Th}(S_i^{ah} \log S_i^{ah} + S_j^{ah} \log S_i^{ah}))$. Since the number of the bitmaps N_b is much smaller than $(S_i^{ah} \log S_i^{ah} + S_j^{ah} \log S_i^{ah})$, the information aggregation with the proposed hybrid sketches greatly reduces the time complexity compared to the intuitive approach. The space required to store the hybrid sketches at peer i is given by $S_{H_i} = b \sum_{h \in H_l} |h| + N_i^{Th} N_b L = b \sum_{h \in H_l} |h| + N_i^{Th} N_b \log_2 S_i^{max-h}$, where S_{H_i} is the space required by the hybrid sketches at peer i, H_l denotes the set of the sketches for the light seeking records, $|h|$ is the number of the seek IDs in hybrid sketch h, and S_i^{max-h} is the maximum number of seek IDs among the heavy seeking records at peer i. If the intuitive approach is used, the space requirement is given by $S_{I_i} = b \sum_{h \in H_l} |h| + b \sum_{h \in H_h} |h|$, where H_h is the set of the sketches for the heavy seeking records. By setting an appropriate threshold γ_{ID}, we can have $N_b \log_2 S_i^{max-h} < b|h|, \forall h \in H_h$. Therefore $S_{H_i} < S_{I_i}$. In other words, the hybrid sketches require a much smaller space compared to the intuitive approach.

4.1.2.3 Estimation of Segment Access Probability

After peer i has collected sufficient seeking statistics, it can compute the segment access probability $P(x,y)$. First, peer i estimates the access frequency for each seeking type from the corresponding hybrid sketch. Each seeking type corresponds to a seeking behavior $s(x,y)$. Therefore, we actually have the access frequency for each seeking behavior $s(x,y)$. Let $f(x,y)$ denote the access frequency of seeking behavior $s(x,y)$. The access probability $P(x,y)$ from segment x to segment y is estimated by

$$P(x,y) \approx \frac{f(x,y)}{\sum_{m=1}^{M} \sum_{n=1}^{M} f(m,n)}. \tag{4.2}$$

The conditional access probability at the current segment α is given by

$$P(y|\alpha) = \frac{P(\alpha,y)}{\sum_{y=1}^{M} P(\alpha,y)}. \tag{4.3}$$

4.1.3 Optimal Prefetching Scheme

4.1.3.1 The Optimal Prefetching Scheme

After the segment access probability has been obtained, we can design the optimal prefetching scheme based on the $P(x,y)$. Suppose that the peer is watching segment i currently. After viewing segment i, this peer may seek any other position. In order to have continuous playback, it can prefetch the appropriate segments during the viewing time.

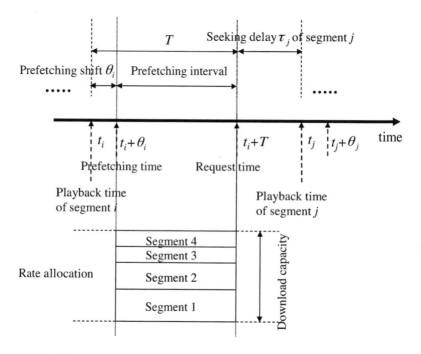

FIGURE 4.2
Determine the prefetching scheme when segment i is being watched

The video has a Constant Bit Rate (CBR) of r_s, and the video is segmented into M segments. The viewing time of a segment is T. Each segment has a fixed size $s_{seg} = r_s T$. A segment is played right after it has been completely downloaded at this peer. We define a whole set Y as the set of all the segments in the video. The prefetching process when the peer is viewing segment i is illustrated in Figure 4.2. The playback time of segment i is denoted by $t_i, \forall i \in Y$. At time t_i, segment i is ready for playback. However, the peer has been prefetching some other segments, which are useful for the future playback. The peer completes these ongoing downloads before it starts the next-round prefetching. So there is a gap between the prefetching time and the playback time of segment i, which is called *prefetching shift* θ_i. After the peer finishes viewing segment i, it will request a new segment (e.g., segment j) at time $t_i + T$. The interval between the request time and the playback time of segment j is denoted as *seeking delay* τ_j of segment j. So the seeking delay is given by $\tau_j = t_j - (t_i + T)$.

Each peer maintains a cache set B in its buffer, and the downloaded segments are put into the cache set. The cached segments help to reduce the seeking delay. If all the segments are cached, the seeking delay is always zero no matter where it jumps. Moreover, the cached segments can be served to the other peers. In P2P applications, each peer is only willing to contribute a limited buffer resource. Therefore a peer can only cache a certain amount of the segments. Let ξ denote the maximum number of the segments in the cache set. The cache set is dynamically updated due to the limited capacity. The cache replacement is completed by every prefetching time. We will describe the cache replacement policy in the following subsection. The cache set during the interval from the prefetching time $(t_i + \theta_i)$ of segment i to the prefetching time of the next segment $(t_j + \theta_j)$ is denoted as B_i. The unavailable set U contains the segments that are not in the cache set. The unavailable set is also varying with time. The unavailable set during the interval from the prefetching time $(t_i + \theta_i)$ of segment i to the prefetching time of the next segment $(t_j + \theta_j)$ is denoted as U_i. So we have $Y = B_i + U_i, \forall i \in Y$.

The download capacity of the receiving peer is denoted by b_d. Each prefetched segment will be downloaded from the serving peers who are caching this segment. The receiving peer can retrieve the desired segment by downloading it from one or multiple serving peers. The upload capacity of segment i provided by the serving peers is denoted as o_i. In P2P VoD applications, upload capacity of segment i is typically less than the download capacity, that is, $o_i < b_d, \forall i \in Y$.

The purpose of the prefetching scheme is to reduce the seeking delay. In the case without prefetching, the peer starts to download the requested segment after the seeking request is generated. Suppose that segment i is being watched, the expected seeking delay without prefetching is given by $E(\tau_i) = \sum_{j \in U} P(j|i) s_{seg}/o_j$. If we prefetch some segments before the request time, the expected seeking delay can be reduced, since some segments in the unavailable set have been downloaded completely or partially before the request time.

The prefetching is performed between the prefetching time of segment i and the request time of segment j; this interval is defined as *prefetching interval*. All the segments in the unavailable set are the prefetching candidates. We allocate each of them a prefetching rate $r_k, \forall k \in U_i$. In P2P applications, the bandwidth bottleneck typically occurs at the access links, therefore we have $r_k \leq o_k$, and $r_k \leq b_d$. We have assumed that $o_k < b_d$, so the constraint $r_k \leq o_k$ covers the both constraints. The minimum required rate to completely download a segment during the prefetching interval is given by $r_m = s_{seg}/(T - \theta_i)$. In order to fully utilize the download bandwidth, we limit the prefetching rate to no larger than r_m, which is $r_k \leq r_m$. Those segments allocated with a rate of r_m will be prefetched completely by the request time $(t_i + T)$ of segment j, while those allocated with a rate less than r_m will be prefetched partially by the request time.

In the prefetching scheme, if the current segment is segment i, we can calculate the expected seeking delay considering two different cases as follows.

1. If the requested segment j is in the cache set B_i, the seeking delay of segment j will be 0. The probability of this case is $P_1 = \sum_{k \in B_i} P(k|i)$.

2. If the requested segment j is in the unavailable set U_i, the prefetched part of segment j during the prefetching interval is $r_j(T - \theta_i)$, the remaining part of segment j will be downloaded at the upload capacity o_j. Therefore, the seeking delay of segment j is given by $\tau_j = \frac{s_{seg} - r_j(T - \theta_i)}{o_j}$.

Then, the expected seeking delay is given by

$$
\begin{aligned}
E(\tau_j) &= \sum_{k \in U_i} \frac{P(k|i)(s_{seg} - r_k(T - \theta_i))}{o_k} \\
&= \sum_{k \in U_i} \frac{P(k|i)s_{seg}}{o_k} - \sum_{k \in U_i} \frac{P(k|i)r_k(T - \theta_i)}{o_k} \\
&= \sum_{k \in Y} \frac{P(k|i)s_{seg}}{o_k} - \sum_{k \in B_i} \frac{P(k|i)s_{seg}}{o_k} - \sum_{k \in U_i} \frac{P(k|i)r_k(T - \theta_i)}{o_k}.
\end{aligned}
$$
(4.4)

The first term $\sum_{k \in Y} \frac{P(k|i)s_{seg}}{o_k}$ in Equation (4.4) represents the expected seeking delay without cache and prefetching. The second term $\sum_{k \in B_i} \frac{P(k|i)s_{seg}}{o_k}$ represents the reduction of the expected seeking delay due to the cache set B_i. The third term $\sum_{k \in U_i} \frac{P(k|i)r_k(T - \theta_i)}{o_k}$ represents the reduction of the expected seeking delay caused by the prefetched part of the segments in the unavailable set U_i.

The *prefetching optimization problem* is to minimize the expected seeking delay by optimally allocating the prefetching rate for each of the segments in the unavailable set. When the current segment is segment i, the optimization problem can be mathematically formulated as follows.

$$\text{minimize}_{\{r_k\}} \quad E(\tau_j) = \sum_{k \in Y} \frac{P(k|i)s_{seg}}{o_k} - \sum_{k \in B_i} \frac{P(k|i)s_{seg}}{o_k}$$
$$- \sum_{k \in U_i} \frac{P(k|i)r_k(T-\theta_i)}{o_k} \qquad (4.5)$$
$$\text{subject to} \quad 0 \leq r_k \leq \min\{o_k, r_m\}, \quad \forall k \in U_i,$$
$$\sum_{k \in U_i} r_k \leq b_d,$$

where r_k is the prefetching rate of the segment k in the unavailable set U_i, o_k is the upload capacity of segment k, r_m is the minimum rate to complete the download of a segment within the prefetching interval, and b_d is the download capacity of the receiving peer.

We define an Access-probability to Upload-capacity Ratio (AUR) as $q_{k|i} = P(k|i)/o_k$, then the prefetching optimization problem is equivalent to the following:

$$\text{maximize}_{\{r_k\}} \quad (T - \theta_i) \sum_{k \in U_i} q_{k|i} r_k$$
$$\text{subject to} \quad 0 \leq r_k \leq \min\{o_k, r_m\}, \quad \forall k \in U_i, \qquad (4.6)$$
$$\sum_{k \in U_i} r_k \leq b_d.$$

The above problem is a Linear Programming (**LP**) problem, which can be solved efficiently using simplex method or interior point method [21]. By solving problem (4.6), we obtain the optimal prefetching rate for each segment in the unavailable set U_i. If a segment is allocated with a zero prefetching rate, it will not be scheduled for prefetching.

Carefully observing problem (4.6), we find that the objective function is a weighted sum of the prefetching rate of each segment in the unavailable set. In order to maximize the objective function, the segment that has a larger AUR $q_{k|i}$ should be allocated with a larger rate within the constrained range. With this in mind, we derive a greedy rate allocation algorithm as follows.

1. Sort the segments in the unavailable set with respect to AUR in descending order;

2. Starting from the segment with the largest value of AUR, allocate as large prefetching rate as possible to each segment in the unavailable set until the download capacity is used up.

We define a prefetched set F_i containing those segments allocated with a positive prefetching rate. In the prefetched set F_i, we categorize those segments allocated with a rate of r_m into a complete set A_i, and the other segments with a rate less than r_m into an ongoing set G_i. We have $F_i = A_i + G_i$. The optimal rate allocation algorithm determines which segments will be prefetched and what the rate is for each prefetched segment. The unavailable set is varying with time, hence the optimal prefetching scheme is performed at every prefetching time.

4.1.3.2   The Optimal Cache Replacement Policy

After completely watching segment i, the peer requests the next segment at request time $t_i + T$. If the requested segment is neither in the current cache set B_i nor in the prefetched set F_i, the peer will download the whole requested segment at its upload capacity. If the requested segment is in the ongoing set G_i, the peer will download the remaining part of the requested segment at its upload capacity. If the requested segment has arrived completely by the request time, the peer starts to play the requested segment right away.

At the request time $t_i + T$, the peer requests the next segment (e.g., segment j). At this time, the peer determines the new cache set B_j. The construction of the new cache set can be formulated as an optimization problem: to maximize the reduction of the expected seeking delay given that the current position is segment j by choosing the optimal segments from the old cache set B_i and the prefetched set F_i, subject to the constraint of the cache capacity ξ. Mathematically, it is formulated as follows.

$$
\begin{aligned}
\text{maximize}_{(B_j)} \quad & \Sigma_{k \in B_j} \frac{P(k|j)s_{seg}}{o_k} \\
\text{subject to} \quad & |B_j| \leq \xi, \\
& B_j \subseteq (B_i + F_i),
\end{aligned}
\tag{4.7}
$$

where B_j is the new cache set, $|B_j|$ denotes the number of the segments in cache set B_j, B_i is the old cache set and F_i is the prefetched set when segment i is being watched.

We can find the optimal solution to problem (4.7) by choosing ξ segments that have the largest value of AUR $q_{k|j}$ from the union set of B_i and F_i to constitute the new cache set B_j. Comparing the new cache set B_j with the old cache set B_i, we know which segments should be ejected from the new cache set, and which segments will be filled into it. Therefore the optimal solution to problem (4.7) actually provides the optimal cache replacement policy. If the newly filled segments are in the ongoing set G_i, the peer needs to complete the download of these segments before starting the next-round prefetching. The chosen ongoing segments will be downloaded completely at time $t_j + \theta_j$, where θ_j is the prefetching shift for segment j. The optimization problem for cache replacement is computed at time $t_i + T$. However, the new cache set B_j is not formed until the prefetching time $t_j + \theta_j$.

4.1.4   Simulation Results

In the simulations, we choose a video clip of 30 minutes. The video is evenly divided into 90 segments. Each segment has a fixed duration of 20 seconds. We randomly generate the seeking behaviors of 1000 sessions, in which users seek more frequently to the popular segments. The peers participating in this video session collect the seeking statistics from the 1000 sessions and estimate the segment access probability $P(x, y)$, which is used to guide their seeking behaviors. In the default setting, the cache capacity is 10 segments, the upload

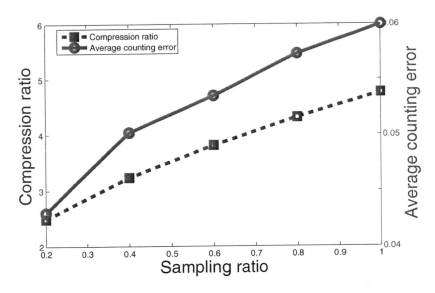

FIGURE 4.3
Comparison of the compression ratio and the average counting error at different sampling ratios

capacity for each segment is uniformly distributed between 500 Kbps and 800 Kbps, and the download capacity of the receiving peer is 1600 Kbps. The playback rate of the video is 500 Kbps. In the hybrid sketches for collecting the seeking statistics, we set the threshold γ_{ID} to 4. If the number of seek IDs in a seeking record is not larger than 4, the hybrid sketch is the enumeration of all the seek IDs. Otherwise, the hybrid sketch is an FM sketch, which consists of 4 bitmaps, each having a length of 11 bits.

4.1.4.1 Performance Evaluation of Seeking Statistics Aggregation

We first evaluate the performance of seeking statistics aggregation by comparing the proposed hybrid sketches with the intuitive approach. The receiving peer can collect a part or all of the seeking statistics from the previous 1000 sessions to estimate the segment access probability $P(x, y)$. We define a sampling ratio $\beta = N_P^S / N_T^S$, where N_P^S is the number of the seek IDs that have been collected, and N_T^S is the number of the total seek IDs in the 1000 sessions. We evaluate the compression ratio and the counting error with different sampling ratios in Figure 4.3. The compression ratio is defined as $\lambda_\beta = S_I^\beta / S_H^\beta$, where S_I^β is the space size using the intuitive approach at sampling ratio β, and S_H^β is the space size using the hybrid sketches at sampling ratio β. When the seeking statistics are represented in the proposed hybrid sketches, the space requirement for a seeking record is fixed at 44 bits when the number of the elements in the seeking record is larger than 4. On the other hand, the space

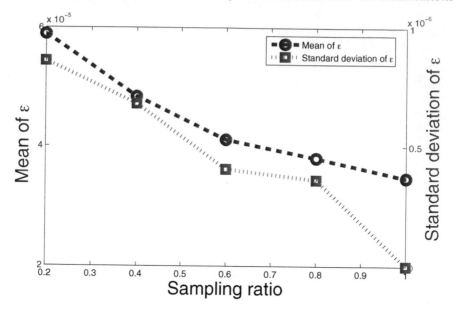

FIGURE 4.4

Comparison of the mean absolute error ε at different sampling ratios

requirement for a seeking record is linearly increasing with the number of the seek IDs in the intuitive approach. Therefore, we can see in Figure 4.3(a) that a larger sampling ratio leads to a larger compression ratio. The counting error for a seeking type is defined as $e_m^\beta = |f_m^{I-\beta} - f_m^{H-\beta}|/f_m^{I-\beta}$, where $f_m^{I-\beta}$ is the access frequency of seeking type m counted with the intuitive approach at sampling ratio β, and $f_m^{H-\beta}$ is the access frequency of seeking type m estimated from the proposed hybrid sketch. The intuitive approach can obtain the accurate count for each seeking type. When the access frequency of a seeking type is not larger than 4, the hybrid sketch use the same counting method as the intuitive approach, hence the counting error is 0. When the access frequency of a seeking type is larger than 4, the hybrid sketch uses FM sketch to estimate the frequency, which introduces an error. When the sampling ratio is increased, a larger portion of seeking types has an access frequency larger than 4, which means FM sketches take a larger portion in all the hybrid sketches. This increases the average counting error, as shown in Figure 4.3(b). Even so, the maximum average counting error with full sampling is still very small (e.g., less than 6.0%). The proposed hybrid sketches introduce a small counting error in counting the access frequency of each seeking type, which is tolerable in the applications of user behavior prediction.

We predict the future user behaviors from the previous seeking statistics. Therefore, the more seeking statistics we collect, the more accurately we can estimate the access probability. In the simulation, we define a *good prediction*

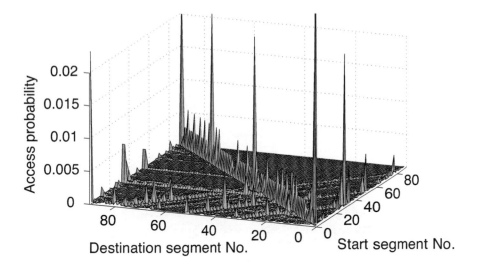

(a)

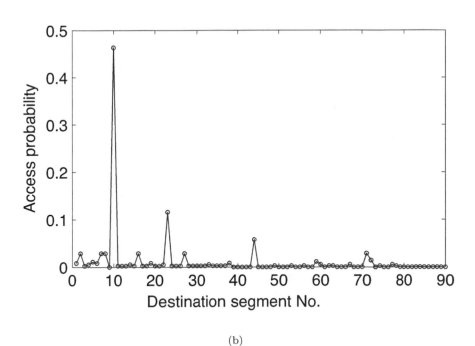

(b)

FIGURE 4.5
Segment access probability estimated from FM sketches with full sampling, (a)
2-D segment access probability $P(x, y)$, and (b) conditional access probability
given that the current position is segment 9

as the prediction of segment access probability from all the seeking statistics in the previous 1000 sessions. The collection of only partial seeking statistics leads to a deviation from the good prediction. We use Mean Absolute Error (MAE) to measure this deviation. The MAE at sampling ratio β is defined as $\varepsilon_\beta = \frac{1}{N^T} \sum_{k=1}^{N^T} |P_k^{H-\beta} - P_k^{I-\beta}|$, where N^T is the number of the seeking types, $P_k^{H-\beta}$ is the segment access probability for seeking type k estimated from a hybrid sketch at sampling ratio β, and $P_k^{I-\beta}$ is the segment access probability for seeking type k computed from the intuitive approach with full sampling. We run 10 simulations for each sampling ratio β, and then calculate the mean and standard deviation of the MAEs. At a smaller sampling ratio, a shorter time is required to complete the statistics aggregation. However, the mean MAE is larger, as shown in Figure 4.4(a), meaning that collecting a smaller amount of seeking statistics causes a larger deviation from the good prediction. In addition, each peer may collect different statistics, thus leading to a different access probability. This is reflected in the standard deviation of MAEs shown in Figure 4.4(b). When each peer collects all the statistics, all of them will have a common segment access probability; in that case the standard deviation of MAEs is 0.

The segment access probability estimated from the hybrid sketches with full sampling is shown in Figure 4.5(a). The MAE in this case is 3.47×10^{-5}. The small MAE indicates that the estimated access probability is close to the accurate one. We observe that most of the peaks appear in the $(x, x + 1)$ positions, indicating that the peer does sequential playback at most of the moments. We also show the conditional access probability given that the current position is segment 9 in Figure 4.5(b).

4.1.4.2 Performance Evaluation of the Prefetching Scheme

We compare the seeking delay in two prefetching schemes: 1) the proposed prefetching scheme, as described in Section 4.1.3; and 2) the sequential prefetching scheme, in which the peer prefetches the segments next to the current segment sequentially, and the segments closest to the current viewing position are cached in priority.

We generate 50 sessions, in which the peer performs guided seeks based on the segment access probability learned from the previous seeking statistics with full sampling. Please note that the sequential playback is a special seek from the current segment x to segment $x + 1$. The initial cache is empty, and the downloaded segments are cached until the cache buffer reaches its capacity; then the cache replacement is performed to maintain the cache level at its capacity. The seeking delay for each seek in an individual session is shown in Figure 4.6(a), we observe that the proposed prefetching scheme achieves zero seeking delay for 84% of the total seeks, higher than the sequential prefetching scheme by 33%. Figure 4.6(b) shows the average seeking delay among 50 sessions. The proposed scheme reduces the average seeking delay by 4.1 seconds in average over the sequential scheme.

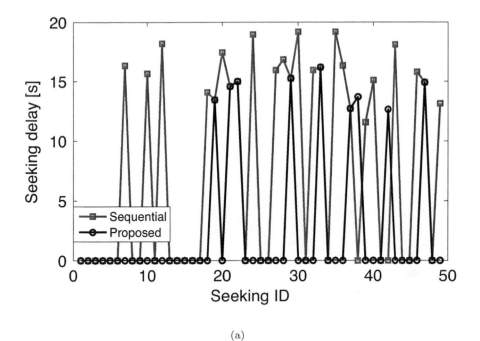

(a)

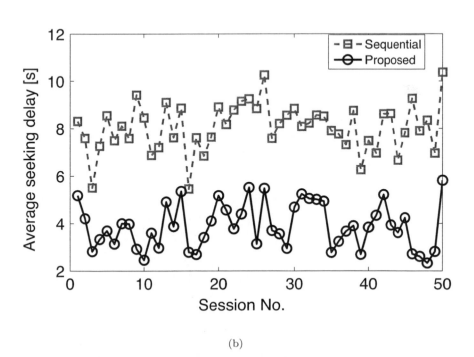

(b)

FIGURE 4.6
Comparison of the seeking delay: (a) the seeking delays in an individual session, and (b) the average seeking delay in 50 sessions

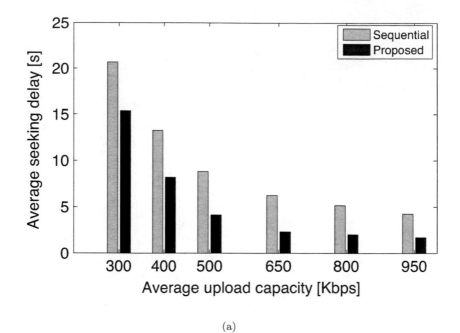

(a)

(b)

FIGURE 4.7
Comparison of the average seeking delay in a session by varying: (a) the average upload capacity, and (b) the download capacity of the receiving peer

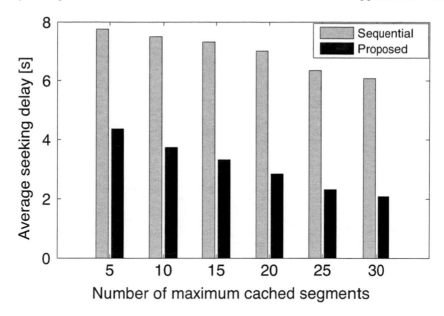

FIGURE 4.8
Comparison of the average seeking delay in an individual session with a different number of maximum cached segments

The seeking delay is dependent on both the upload capacity and download capacity. In the P2P VoD applications, the segment is served by one or multiple serving peers who are buffering it. Therefore the upload capacity of a segment is different from segment to segment. We vary the average upload capacity from 300 Kbps to 950 Kbps, and compare the average seeking delay in two prefetching schemes in Figure 4.7(a). When the upload capacity is increased, the average seeking delay is reduced in both schemes. The proposed prefetching scheme outperforms the sequential scheme by an average of 4.11 seconds in the average seeking delay. A larger download capacity allows the peer to prefetch more segments, thus reducing the seeking delay. As seen in Figure 4.7(b), the seeking delay in both schemes is reduced when the download capacity is increased from 800 Kbps to 2400 Kbps. Among different download capacities, the seeking delay in the proposed prefetching scheme is 2.5–4.1 seconds smaller than the sequential prefetching scheme.

We show the impact of the cache set on the average seeking delay in Figure 4.8. As we increase the number of the maximum cached segments, we can achieve a smaller seeking delay in both schemes. When the cache capacity is increased from 5 to 30 segments, the average seeking delay is reduced by 2.1 seconds in the sequential prefetching scheme and 1.5 seconds in the proposed scheme. Among different cache capacities, the proposed prefetching

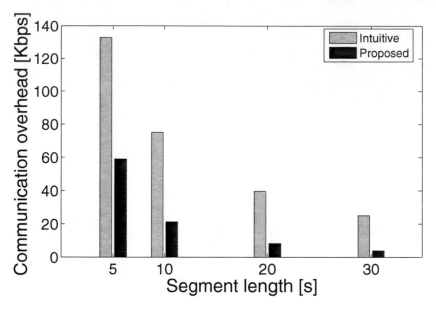

FIGURE 4.9
Comparison of the average communication overhead with different segment
length

scheme improves the seeking delay by average 3.9 seconds over the sequential
prefetching scheme.

We also perform experiments to compare the communication overhead
with different segment length in Fig. 4.9. The communication overhead in
collecting the seeking statistics depends on the number of the seeking types
and the sampling period. The sampling period is defined as the time interval
between two consecutive gossips. A smaller segment length leads to a larger
communication overhead because 1) it produces a dramatically large number
of seeking types, and 2) it requires a smaller sampling period in order to catch
up with the faster prefetching pace. When the segment length is increased
from 5 seconds to 30 seconds, the communication overhead used in collecting the seeking statistics with the proposed hybrid sketches is reduced from
59.0 Kbps to 4.0 Kbps. From Fig. 4.9, we can see that the proposed hybrid
sketches greatly reduce the communication overhead compared to the intuitive
approach.

4.1.5   Practical Issues

In P2P VoD applications with guided seeks, users seek to different segments
frequently. It is critical for the peer to quickly locate the destination segment.
In our framework, we can use information-assisted peer search to quickly find

the serving peers. This approach requires each peer to maintain a segment-location table, which consists of entries of three-component tuples (the segment no., IP address, and the update time). Each peer aggregates the segment location information when it collects the seeking statistics via gossips. The information exchange process is as follows. First, peer i updates the segment location table based on its cache set. Second, peer i exchanges the location information with its randomly chosen neighbors in a pair-wise way. In each pair-wise exchange, peer i checks each segment entry in the segment-location table of its neighbor, and insert it into its segment-location table if the segment entry has not existed in its table, or the segment entry has a recent update time.

After peer i determines the segments for prefetching, it needs to quickly locate the serving peers. We use information-assisted random walks to find the segment owners. Peer i first looks them up in its segment-location table. If the serving peers are found, peer i will request the segments from them. Otherwise, peer i sends the query to one or multiple of its neighbors, who will check its segment-location table and feed back the locations to peer i if it is found. With the assistance of segment-location tables, the desired segment is expected to be located within fewer hops compared to the random walks without information assistance [22], thus greatly reducing the search delay and the search overhead.

With the hybrid sketches, the maximum space requirement is $M(M-1)N_bL$. The space size is not increasing with the number of seeking statistics. Therefore we can collect as many as possible seeking statistics. The more seeking statistics we collect, the more accurate segment access probability we can obtain.

In P2P VoD applications, one of the practical problems is the peer dynamics (e.g., random join, departure or failure). In the proposed framework, we collect the seeking statistics and segment location information using gossip protocol, which has been proven to be robust to the peer dynamics [18]. If a peer leaves the P2P network, the information (e.g., seeking statistics and segment-location information) at this peer has been spread to the other peers prior to its departure with a high probability. When a new peer joins the P2P network, it can quickly aggregate the information by gossiping with other peers who have collected a lot of information.

Each prefetched segment is uploaded by the serving peer who is buffering this segment. The serving peer of a segment is changing with time, since 1) the cached segment may be replaced by another newly downloaded segment, and 2) a peer may leave or join the P2P network at any time. Therefore, we need to estimate the upload capacity of each segment dynamically. With periodic gossip, we can collect the statistics of the upload rate for a segment in the recent past, and then predict the upload capacity of this segment in the near future.

4.2   Optimization of Substream Allocation in Layered P2P Applications

In P2P streaming, peers communicate directly with each other for sharing and exchanging data as well as other resources such as storage. It is typical that several serving peers collectively stream the requested content to the receiving peer, since a single serving peer may not be able or willing to contribute an upload bandwidth enough for the media playback at the receiver. This multipoint-to-point communication can provide a higher overall throughput to the receiver, hence resulting in a higher quality [23].

In layered coded P2P streaming, the video is encoded into multiple layers. A layer will not be able to be decoded if the lower layers are not available. The bit stream in layer i is defined as substream i. After a streaming session is finished, the receiving peer will replicate one or multiple substreams, and become a qualified peer [24], which may serve other requesting peers in the next streaming session.

One of the major challenges in the layered P2P systems is how to allocate different numbers of the substreams for different layers. The lower layers have more importance than the higher layers. Therefore, more replicas should be made, to increase the availability of the lower layers. The substream allocation is implemented by substream replication in the pool of qualified peers. A good allocation scheme can help the system to provide an overall good quality to the users, and accelerate the growth of the peer population in the P2P systems.

The impact on the system performance with different allocation schemes has been studied in [25]. In this section, we propose a novel allocation scheme in layered P2P streaming systems. Compared with the general allocation schemes with fixed allocation percentages [25], the proposed scheme can provide the users with a higher quality. We also study how the proposed allocation scheme can accelerate the growth of the peer population in the initial stage of hybrid P2P streaming systems.

4.2.1   Optimal Substream Allocation Scheme

4.2.1.1   Single-File system

We first study a layered P2P streaming system with only a single video k where k is the video ID. In this subsection, we would like to determine how many copies of each substream should be stored among the pool of the qualified peers in order to achieve an overall high quality, if the total number of the online qualified peers N_k is given.

The mean duration of the video is L. We encode the video into M layers using layered video codec. Each layer has an average bit rate B_r. We assume that each substream is delivered from a peer via an independent path. The probability of successfully receiving substream i is denoted as q_{ki}.

In our model, a central server manages the qualified peers and controls the substream replication. Each request is first sent to the central server, which then selects a qualified peer to deliver substream $i(i = 1, 2, ..., M)$ to the receiver, respectively.

We assume that each qualified peer is homogeneous in contributed storage and upload bandwidth. Each qualified peer can store only one substream and upload only one substream to the receiver. Each qualified peer alternates between "on" and "off" states. We model the "on" time as an exponentially distributed random variable with mean T_{on}, and the "off" time as another exponentially distributed random variable with mean T_{off} [25]. The steady probability that a peer stays at the "on" state is given by $u = T_{on}/(T_{on}+T_{off})$. There is a total of W_k qualified peers for video k in the pool. The expected number of the online qualified peers is $N_k = uW_k$. We will next determine the number N_{ki} of substream $i(i = 1, 2, ..., M)$ in the pool in order to achieve a high expected quality.

The expected quality is related to the request rate and the number of the substream for each layer. The request to the video is modeled as a Poisson process with a rate λ_k. The qualified peers with substream i are trunked together to serve the requests for substream i. A request will be rejected without waiting in a queue, if there is no corresponding substream available in the serving pool. The steady probability P_{Bki} that a request for substream i is rejected can be found from the Erlang B trunking model [24, 18]:

$$P_{Bki} = \frac{(\lambda_k L)^{N_{ki}} \frac{1}{(N_{ki})!}}{\sum_{n=0}^{N_{ki}} \frac{(\lambda_k L)^n}{n!}}. \tag{4.8}$$

We assume that the session will be finished successfully once its request is accepted. The probability of receiving substream i can be given by $q_{ki} = 1 - P_{Bki}$. The probability P_{km} that the user receives the video with m layers can be written as

$$P_{km} = \begin{cases} 1 - q_{k1}, & \text{if } m = 0, \\ (1 - q_{k(m+1)}) \prod_{i=1}^{m} q_{ki}, & \text{if } m = 1, 2, ..., (M-1), \\ \prod_{i=1}^{M} q_{ki}, & \text{if } m = M. \end{cases} \tag{4.9}$$

The expected distortion of the received video is given by

$$E(D_{kM}) = \sum_{m=0}^{M} P_{km} D_{km}, \tag{4.10}$$

where D_{km} is the average distortion of received video with m decodable layers.

The proposed substream allocation scheme is illustrated in Figure 4.10. The expected number of the online qualified peers N_k is equal to the number of the substreams stored in the qualified peers, since each peer stores only one substream in our model. We determine the number of substream i using a greedy algorithm as follows.

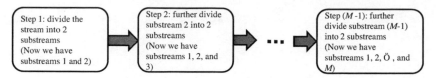

FIGURE 4.10

Steps of the proposed substream allocation scheme

At step 1, we divide N_k substreams into two kinds: substream 1 and substream 2. We then find the number N_{k1}^* of substream 1 by searching in N_k in order to minimize the expected distortion $E(D_{k2})$, the expected distortion of received video with at most 2 layers. At step $i(i = 2, 3, ..., (M - 1))$, given $N_{kj}^*(j = 1, 2, ..., (i - 1))$, we divide N_k into $(i + 1)$ layers and search within the number of the remaining substreams $N_L(i) = N_k - (N_{k1}^* + N_{k2}^* +, ..., + N_{k(i-1)}^*)$ to find N_{ki}^* in order to

$$\text{maximize}_{(N_{ki}^*)} \quad E(D_{k(i+1)}) = \sum_{m=0}^{i+1} P_{km} D_{km}$$
$$\text{subject to} \quad \sum_{j=1}^{i+1} N_{kj}^* = N_k. \tag{4.11}$$

At step $(M - 1)$, we can obtain $N_{k(M-1)}^*$. The left substreams are assigned to layer M: $N_{kM}^* = N_k - (N_{k1}^* + N_{k2}^* +, ..., + N_{k(M-1)}^*)$. Finally, we obtain the number of the substreams for each of the M layers, which can be denoted with an allocation vector $V_k = \{N_{k1}^*, N_{k2}^*, ..., N_{kM}^*\}$.

The proposed scheme allocates the number of the copies for each substream by adapting to the request rate and the total number of substreams. The central server carries out the substream allocation by controlling the substream replication in each qualified peer, such that all the substreams are placed following the allocation vector V_k.

4.2.1.2 Multi-File system

We now extend the analysis to a simplified multi-file system with F files. Each file has an average length of L. We divide the multi-file system into F virtual sub-systems, each of which deals with only one file.

The popularity of video $k(k = 1, 2, ..., F)$ is modeled as a random variable following $Zipf$ distribution $f_Z(k)$. The aggregate request rate is λ_T. The request rate to video k is $\lambda_k = \lambda_T f_Z(k)$. There are totally N_T online qualified peers in the system; we allocate N_k qualified peers to store video k based on its popularity. Therefore we have $N_k = N_T f_Z(k)$.

Given the number of the online qualified peers for video k, we then can further allocate the substreams for each layer of video k using the proposed allocation scheme. The substream allocation vector for video k can be obtained as $V_k = \{N_{k1}^*, N_{k2}^*, ..., N_{kM}^*\}$.

The average distortion of the F videos in the simplified multi-file system is given by

$$D_T = \sum_{k=0}^{F} f_Z(k)E(D_{kM}),\qquad(4.12)$$

where $E(D_{kM})$ is the expected distortion of video k with at most M layers.

4.2.2   Evolution of Video Quality and Peer Population

In this subsection, we study how the expected video quality and peer population evolve under the proposed substream allocation scheme in the hybrid P2P streaming. In a hybrid P2P system, a newly released video file is first pushed to the central server. In the initial stage, the central server cooperates with the qualified peers to serve streams to the users, since there are insufficient qualified peers available [24].

In our model, each qualified peer stores and uploads only one substream. We assume the Poisson request rate is fixed at λ_H. The length of the video is L. We analyze the population of the qualified peers in a discrete-time manner. Each interval has a length of L. If G streaming sessions are initiated in the $n^{th}(n = 1, 2, ...)$ interval and completed successfully, these G receivers will become new qualified peers in the $(n + 1)^{th}$ interval [24].

The server has a fixed upload bandwidth, which can deliver S substreams. The population of the qualified peers in the nth is denoted as $Z(n)$, which is equal to the number of the substreams in the nth interval, since each qualified peer stores one substream. Under the proposed allocation scheme, the total $(S + Z(n))$ substreams are deployed as $V_H(n) = \{N_1(n), N_2(n), ..., N_M(n)\}$, where M is the number of the layers, $N_i(n)(i = 1, 2, ..., M)$ is the number of the substream i in the nth interval. Given the allocated number of each substream and the request rate λ_H, the expected distortion of the received video in the nth interval can be calculated from Equation (4.10).

In the $(n+1)^{th}$ interval, the expected number of newly generated qualified peers $Y_m(n + 1)$ with m decodable layers is given by

$$Y_m(n + 1) = \lambda_H L P_m(n), \quad m = 1, 2, ..., M,\qquad(4.13)$$

where $P_m(n)$ is the probability that a peer receives a video with m decodable layers in the nth interval. The population of the qualified peers in $(n + 1)^{th}$ interval is given by $Z(n+1) = Z(n)+(Y_1(n+1)+Y_2(n+1)+, ..., +Y_M(n+1))$. Under the proposed allocation scheme, the total $(S + Z(n + 1))$ substreams in the $(n + 1)^{th}$ interval are deployed as $V_H(n + 1) = \{N_1(n + 1), N_2(n + 1), ..., N_M(n + 1)\}$. Then the expected distortion of the received video in the $(n+1)^{th}$ interval can also be calculated from Equation (4.10). In this way, we can get the evolution of the expected quality and expected population of the qualified peers under the proposed allocation scheme.

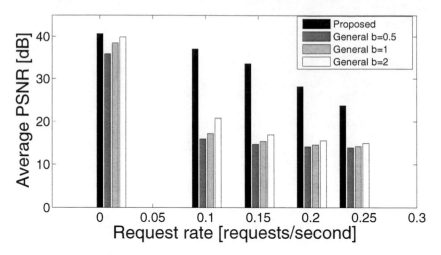

FIGURE 4.11
Average PSNR with varying request rate

4.2.3   Simulation Results

We encode the "Akiyo" Common Intermediate Format (CIF) sequence into 6 layers using SNR-scalable extension of H.264/AVC [14]. Each layer has an average bit rate of 25 Kbps. The total average bit rate of the video is 150 Kbps. Each Group of Pictures (GOP) consists of 16 frames. There are 20 videos in the multi-file P2P system. The popularity of the video follows $Zipf$ distribution with parameter $\alpha = 1.1$. The number of the requests follows Poisson process with a rate λ ranging from 0.001 to 0.25 requests/s. The average duration of the video is 1 hour. Each peer contributes an upload bandwidth of 30 Kbps, and an average storage of 15 Mbytes. The mean "on" time of the peer is 9 hours, and the mean "off" time is 1 hour. Therefore, the availability of the peer is 0.9. We compare the proposed allocation scheme to three general allocation schemes [25], in which the number of the copies C_m of layer $m(m = 1, 2, ..., M)$ is given by $C_m = a(M - m + 1)^b$, where a is a normalization constant, b is 0.5, 1.0, or 2.0, respectively [25].

Figure 4.11 shows the average Peak Signal-to-Noise Ratio (PSNR) versus request rate under different allocation schemes. The number of the qualified peers is 1000. The general schemes allocate the channels with fixed percentages without adapting to the change of the request rates, therefore the expected quality drops dramatically when the request rate is large. On the other hand, the proposed scheme adapts to the request rate, thus achieving a higher quality. Figure 4.12 shows the average PSNR with different number of the qualified peers when the request rate is fixed at 0.1 requests/s. When the number of the qualified peers is small, the proposed allocation scheme allocates a larger percentage of the substreams to the lower-layers, thus increasing the qual-

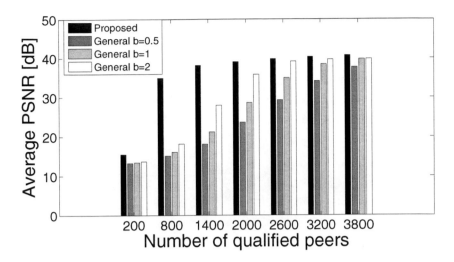

FIGURE 4.12
Average PSNR with varying number of qualified peers

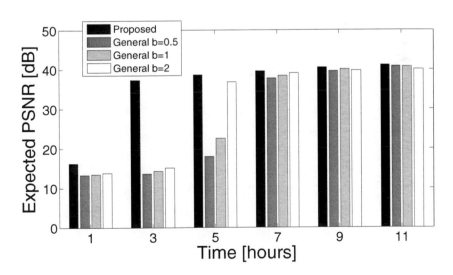

FIGURE 4.13
Evolution of expected PSNR

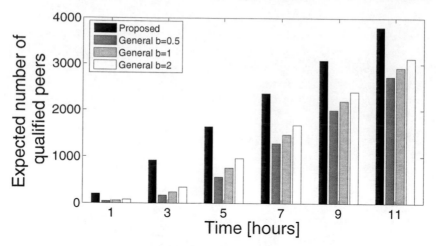

FIGURE 4.14
Evolution of expected number of qualified peers

ity quickly, while the general schemes do not allocate the substreams in an efficient way, thus lowering the quality.

The evolution of expected quality and the number of the qualified peers in the initial stage of hybrid P2P streaming system is evaluated in Figure 4.13 and Figure 4.14, respectively. The upload bandwidth that the server uses to serve this video is 5 Mbps, which can support 200 concurrent substreams. The request rate is 0.1 requests/s. We do not consider peer-failure in this scenario. Figure 4.13 shows average quality evolution with time. At the beginning, the proposed scheme allocates more lower-layer substreams to reduce the number of the rejected requests. Once the number of lower-layer substreams is large enough, the system will increase the number of higher-layer substreams to enhance the quality. The general schemes allocate the number of substreams for each layer based on a fixed percentage. However, most of the high-layer substreams cannot be decoded in the initial stage since the availability of the lower-layer substreams is low. The similar impact can be seen on the peer population growth shown in Figure 4.14. The proposed scheme allocates the substreams in a better way, such that it can generate more qualified peers.

4.3   Summary

In this chapter, we have presented an optimal prefetching framework to reduce the seeking delay in P2P VoD applications. We introduced the concept of guided seeks. With the guidance, users can efficiently seek to the exciting or interesting segments. The guidance information is learned from the seeking statistics in the previous and/or concurrent sessions. It is a nontrivial task to aggregate the seeking statistics in distributed P2P networks. We designed the hybrid sketches to represent the seeking statistics, thus eliminating the double-counting problem, and greatly reducing the space and time complexity. After aggregating the seeking statistics, we estimated the segment access probability, based on which we developed an optimal prefetching scheme and an optimal cache replacement policy to minimize the expected seeking delay at every viewing position.

We have also presented an optimal substream allocation scheme in layered P2P VoD applications. The number of substreams for each layer is determined by a greedy algorithm. The proposed allocation scheme enables layered P2P systems to achieve an overall higher quality compared to the fixed-percentage allocation schemes. In addition, the proposed allocation scheme can accelerate the growth of the video quality and the peer population in the initial stage of hybrid P2P streaming systems.

Bibliography

[1] J. Li, "PeerStreaming: An On-Demand Peer-to-Peer Media Streaming Solution Based On A Receiver-Driven Streaming Protocol," *Proc. of IEEE MMSP,* pp. 1-4, Oct. 2005.

[2] M. Hefeeda, A. Habib, B. Botev, D. Xu, and B. Bhargava, "PROMISE: Peer-to-Peer Media Streaming Using CollectCast," *Proc. of ACM Multimedia,* pp. 45-54, Nov. 2003.

[3] C. Zheng, G. Shen, and S. Li, "Distributed Prefetching Scheme for Random Seek Support in Peer-to-Peer Streaming Applications," *Proc. of ACM MM,* pp. 29-38, Nov. 2005.

[4] A. Sharma, A. Bestavros, and I. Matta, "dPAM: a distributed prefetching protocol for scalable asynchronous multicast in P2P systems," *Proc. of IEEE INFOCOM,* vol. 2, pp. 1139-1150, Mar. 2005.

[5] Y. Shen, Z. Liu, S. Panwar, K.W. Ross, and Y. Wang, "On the Design of Prefetching Strategies in a Peer-Driven Video-on-Demand System," *Proc. of IEEE ICME,* pp. 817-820, Jul. 2006.

[6] Y. He, G. Shen, Y. Xiong, and L. Guan, "Optimal prefetching scheme in P2P VoD applications with guided seeks," *IEEE Transactions on Multimedia,* vol. 11, no. 1, pp. 138-151, Jan. 2009.

[7] T. T. Do, K. A. Hua, and M. A. Tantaoui, "P2VoD: Providing Fault Tolerant Video-on-Demand Streaming in Peer-to-Peer Environment," in *Proc. of IEEE ICC,* vol. 25, no. 1, pp. 119-130, Jan. 2004.

[8] Y. Cui, B. Li, and K. Nahrstedt, "oStream: asynchronous streaming multicast," *IEEE Journal on Selected Areas in Communications,* vol. 22, no. 1, pp. 91-106, Jan. 2004.

[9] Z. Li and A. Mahanti, "A Progressive Flow Auction Approach for Low-Cost On-Demand P2P Media Streaming," *Proc. of QShine,* Aug. 2006.

[10] C. Huang, J. Li, and K. W. Ross, "Peer-Assisted VoD: Making Internet Video Distribution Cheap," *Proc. of IPTPS,* Feb. 2007.

[11] W. P. Yiu, X. Jin and S. H. Chan, "VMesh: Distributed segment storage for peer-to-peer interactive video streaming," *IEEE Journal on Selected Areas in Communications,* vol. 25, no. 9, pp. 1717-1731, Dec. 2007.

[12] X. Xu, Y. Wang, S. P. Panwar, and K. W. Ross, "A Peer-to-Peer Video-on-Demand System using Multiple Description Coding and Server Diversity," *Proc. of IEEE ICIP,* vol. 3, pp. 1759-1762, Oct. 2004.

[13] B. Cheng, H. Jin, X. Liao, "Supporting VCR Functions in P2P VoD Services Using Ring-Assisted Overlays," *Proc. of IEEE ICC,* pp. 1698-1703, Jun. 2007.

[14] K. Suh, C. Diot, J. Kurose, L. Massoulie, C. Neumann, D. Towsle, and M. Varvello, "Push-to-peer video-on-demand system: design and evaluation," *IEEE Journal on Selected Areas in Communications,* vol. 25, no. 9, pp. 1706-1716, Dec. 2007.

[15] J. Kangasharju, K. W. Ross, and D. A. Turner, "Optimizing File Availability in Peer-to-Peer Content Distribution," *Proc. of IEEE INFOCOM,* pp. 1973-1981, May 2007.

[16] N. J. Tuah, M. K., S. Venkatesh, and S. K. Das, "Performance Optimization Problem in Speculative Prefetching," *IEEE Transactions on Parallel and Distributed Systems,* vol. 13, no. 5, pp. 471-484, May 2002.

[17] M. Angermann, "Analysis of Speculative Prefetching," *ACM SIGMOBILE Mobile Computing and Communications Review,* vol. 6, no. 2, pp. 13-17, Apr. 2002.

[18] X. Zhang, J. Liu, B. Li, and T. P. Yum, "CoolStreaming/DONet: A Data-Driven Overlay Network for Efficient Live Media Streaming," *Proc. of IEEE INFOCOM,* vol. 3, pp. 2102-2111, Mar. 2005.

[19] S. Boyd, A. Ghosh, B. Prabhakar, and D. Shah, "Gossip Algorithms: Design, Analysis, and Applications," *Proc. of IEEE INFOCOM,* vol. 3, pp. 1653-1664, Mar. 2005.

[20] P. Flajolet and G. N. Martin, "Probabilistic counting algorithms for data base applications," *Journal of Computer and System Sciences,* vol. 31, no. 2, pp. 182-209, 1985.

[21] R. J. Vanderbei, *Linear programming: foundations and extensions,* 2nd Edition, Springer Press, 2001.

[22] C. Gkantsidis, M. Mihail, and A. Saberi, "Random Walks in Peer-to-Peer Networks," *Proc. of IEEE INFOCOM,* vol. 1, pp. 120-130, Mar. 2004.

[23] D. Xu, M. Hefeeda, S. Hambrusch, and B. Bhargava, "On peer-to-peer media streaming," in *Proc. of IEEE ICDCS,* pp. 363-371, Jul. 2002.

[24] Y. Tu, J. Sun, M. Hefeeda, Y. Xia, S. Prabhakar, "An Analytical Study of Peer-to-Peer Media Streaming Systems," *ACM Transactions on Multimedia Computing, Communications,* vol. 1, no. 4, pp. 354-376, Nov. 2005.

[25] Y. Shen, Z. Liu, S. S. Panwar, K. W. Ross, and Y. Wang, "Streaming Layered Encoded Video Using Peers," in *Proc. of ICME,* Jul. 2005.

[26] T. S. Rappaport, *Wireless Communications: Principles and Practice,* 2nd edition, Prentice Hall, 2002.

[27] H. Schwarz, D. Marpe, and T. Wiegand, "SNR-scalable extension of H.264/AVC," in *Proc. of IEEE ICIP,* vol. 5, pp. 3113-3116, Oct. 2004.

[28] Y. He, I. Lee, and L. Guan, "Substream allocation in layered P2P streaming," in *Proc. of IEEE ICME,* pp. 1505-1508, Jul. 2006.

5

Video Streaming over Wireless Ad Hoc Networks

CONTENTS

Wireless ad hoc networks consist of a collection of wireless nodes which dynamically exchange data among themselves. Recently there is a compelling need to support real-time video streaming over wireless ad hoc networks. Dependent on the number of the simultaneous receivers, video streaming can be classified into unicast streaming and multicast streaming [1]. In this chapter, we examine both unicast streaming and multicast streaming over wireless ad hoc networks.

In the first part of this chapter, we present the system models. We first describe the video distortion model to capture the rate distortion relationship. We then use a network graph to model the network topology. Finally we model the packet loss process at each link with a two state Markov chain, and derive the end-to-end packet loss rate for each path.

In the second part of this chapter, we examine the video unicast streaming over wireless ad hoc networks. Specifically, we propose a fully distributed algorithm to jointly optimize the source rate and routing scheme. Simulation results show that the proposed routing scheme outperforms the existing multi-path routing schemes.

In the third part of this chapter, we extend our study to the video multicast streaming over wireless ad hoc networks with Gaussian broadcast channels using Frequency Division Multiple Access (FDMA). We maximize the aggregate throughput at all receivers by jointly optimizing both the source rate allocation and the routing scheme. Simulation results show that the proposed video multicast scheme yields a superior video quality compared to the double-tree routing scheme.

In the fourth part of this chapter, we take into account the power allocation in the physical layer, and propose a distributed algorithm to jointly optimize the source rate, the routing scheme and the power allocation for video multicasting over wireless ad hoc networks with Code Division Multiple Access (CDMA). We use hierarchical dual decompositions to separate the optimization problem into multiple subproblems, which are then solved in parallel. Through extensive simulations, we demonstrate that the proposed video multicast scheme can achieve much higher video quality compared to both the uniform-power scheme and the tree-based routing schemes.

5.1   System Models

5.1.1   Video Distortion Model

We employ the prioritized coding scheme [2] in the applications of video streaming over wireless ad hoc networks. With a prioritized coding scheme, the expected distortion of the reconstructed video is determined by the expected throughput. The higher throughput a receiver receives, the higher the quality it can reconstruct. A rate-distortion model for single-layered video transmission is reported in [3]. We extend this model to characterize the rate-distortion relationship for the prioritized coded video with protection redundancy. It is given by [2]

$$d = D_0 + \frac{\theta_0}{R + \phi_0}, \tag{5.1}$$

where d is the expected distortion of the reconstructed video, R is the expected throughput at the receiver, D_0 is the encoding distortion, θ_0 and ϕ_0 are parameters for transmission distortion. The parameters for a specific video sequence can be found using data fitting techniques.

5.1.2 Network Model

We represent the topology of a wireless ad hoc network with a directed graph $\mathbf{G} = (\mathbf{N}, \mathbf{L})$, where $\mathbf{N}$ is the set of wireless nodes and $\mathbf{L}$ is the set of directed wireless links. The nodes can send, receive and relay packets over links. A wireless link is represented as an order pair (i, j) of two distinct nodes that can communicate directly with each other. The network topology can be represented by a node-link incidence matrix $\mathbf{A}$, whose elements are given by

$$a_{il} = \begin{cases} 1, & \text{if } i \text{ is the start node of link } l, \\ -1, & \text{if } i \text{ is the end node of link } l, \\ 0, & \text{otherwise.} \end{cases} \tag{5.2}$$

With network coding, a multicast flow from the source node to H receivers can be viewed as H conceptual unicast sessions [4, 5]. We define $\mathbf{V} = \{1, ..., H\}^T$ as the set of conceptual unicast sessions. Let the conceptual unicast rate at link l for conceptual session h be x_{hl}, the multicast flow rate at link l be y_l, then we have $y_l = \max_{h \in \mathbf{V}} \{x_{hl}\}$, or equivalently $y_l \geq x_{hl}, \forall h \in \mathbf{V}$. For video streaming applications with prioritized coding and network coding, a receiver can reconstruct the video at a low quality even if it does not receive all the encoded packets for a Group of Pictures (GOP). Therefore, we can set the multicast source rate to the maximum value among the H conceptual unicast source rates. Let s_h denote the source rate for conceptual unicast session h, then the multicast source rate s_M is given by $s_M = \max_{h \in \mathbf{V}} \{s_h\}$. The flow conservation holds at each node for each conceptual unicast session, which is expressed by

$$\sum_{l \in \mathbf{L}} a_{il} x_{hl} = \eta_{hi}, \quad \forall h \in \mathbf{V}, \forall i \in \mathbf{N}, \tag{5.3}$$

where η_{hi} is equal to s_h if node i is the source, or equal to $-s_h$ if node i is the receiver for conceptual session h, or 0 for all other cases.

5.1.3 Packet Loss Model

Generally, the communication link is time-varying in a wireless ad hoc network. A wireless link fails when the Signal-to-Interference-and-Noise Ratio (SINR) falls below a given threshold. The failure process of a link can be modeled with a two state Markov chain shown in Figure 5.1. When link l is in GOOD (G) state, the packets transmitted over this link will be received successfully by its downstream node. When link l is in BAD (B) state, the transmitted packets will be lost. The probability from a G state to the next

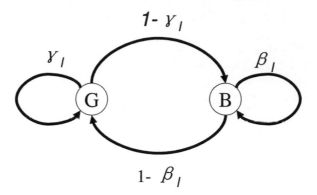

FIGURE 5.1
Two state Markov model

G state is denoted by γ_l, and the probability from a B state to the next B state is denoted by β_l. From the two state Markov chain, we can calculate the packet loss rate p_l at link l as follows.

$$p_l = \frac{1 - \gamma_l}{2 - \gamma_l - \beta_l}, \quad \forall l \in \mathbf{L}. \tag{5.4}$$

For conceptual session h, the source traffic s_h can be split into multiple segments, each traveling through a distinguished single path from the source to the destination. The set of these distinguished paths is denoted by $\mathbf{J_h}$, and the segment traveling through path j_h is denoted by z_j^h. Hence, we have $\sum_{j \in \mathbf{J_h}} z_j^h = s_h$. We define a segment-link matrix E^h, whose elements are given by

$$e_{jl}^h = \begin{cases} 1, & \text{if } z_j^h \text{ travels through link } l, \\ 0, & \text{otherwise.} \end{cases} \tag{5.5}$$

Thus, the link rate x_{hl} is the aggregation of the segments traveling through this link. That is: $x_{hl} = \sum_{j \in \mathbf{J_h}} e_{jl}^h z_j^h$.

The end-to-end packet loss rate that segment z_j^h suffers can be computed as follows.

$$p_{z_j^h}^E = 1 - \prod_{l \in \mathbf{L}} (1 - e_{jl}^h p_l). \tag{5.6}$$

We assume the packet loss rate p_l at each link is small (i.e., $p_l \ll 1$). The end-to-end packet loss rate is thus approximated as

$$p_{z_j^h}^E \approx \sum_{l \in \mathbf{L}} e_{jl}^h p_l. \tag{5.7}$$

The expected end-to-end packet loss rate for conceptual session h is given by

$$p_h^E = \sum_{j \in \mathbf{J_h}} (\frac{z_j^h}{s_h} p_{z_j^h}^E) \approx \sum_{j \in \mathbf{J_h}} (\frac{z_j^h}{s_h} \sum_{l \in \mathbf{L}} e_{jl}^h p_l) = \sum_{l \in \mathbf{L}} p_l \sum_{j \in \mathbf{J_h}} (\frac{z_j^h}{s_h} e_{jl}^h) = \sum_{l \in \mathbf{L}} (p_l \frac{x_{hl}}{s_h}).$$

$$(5.8)$$

The expected video distortion for conceptual session h in the presence of packet loss is given by:

$$
\begin{aligned}
d_h & = D_0 + \frac{\theta_0}{R_h + \phi_0} = D_0 + \frac{\theta_0}{s_h(1 - p_h^E) + \phi_0} \\
& = D_0 + \frac{\theta_0}{s_h(1 - \sum_{l \in \mathbf{L}} (p_l \frac{x_{hl}}{s_h})) + \phi_0} = D_0 + \frac{\theta_0}{(s_h - \sum_{l \in \mathbf{L}} p_l x_{hl}) + \phi_0}.
\end{aligned}
$$

$$(5.9)$$

5.2 Optimized Video Unicasting over Wireless Ad Hoc Networks

With the recent advance in the technologies of digital video and wireless communications, there is an increasing expectation on enabling real-time video streaming over wireless ad hoc networks, in addition to just data communications. For instance, a soccer fan in a stadium may like to use his or her Personal Digital Assistant (PDA) to receive the real-time video of the match that is being played. The service provider can deliver the video from a fixed access point to the subscriber via the relays of other mobile users in the same area.

This section considers the problem of how to deliver a real-time video from a single source to a single receiver over a wireless ad hoc network. We would focus on two issues: routing and source rate allocation. Routing is quite challenging in wireless ad hoc networks due to the dynamic topology and variable channel conditions. In multi-path routing, how to optimally split the traffic into multiple paths is a problem requiring careful investigation. Source rate allocation addresses the problem of how to scale the source rate to meet the available bandwidth of the end-to-end path. Different routing schemes accommodate different optimal source rates.

It has been demonstrated that multi-path transport of multiple video substreams can achieve a significant gain compared to single-path video transmission schemes [6, 7]. Optimization techniques have been used to find the optimal paths. In [14], the authors used Genetic Algorithm (GA) to compute two optimal paths by minimizing the expected distortion. The video was encoded into two descriptions, each transporting over one path. However, GA is computationally intensive and essentially centralized. A distributed algorithm is desired to compute the routing in wireless ad hoc networks. Recently, Zhu et al. have proposed a congestion-minimized routing scheme using a distributed algorithm [9]. However, the link quality (e.g., failure probability) has not yet been considered in [9]. Source rate allocation is often neglected. Previous work [6, 7, 14] simply allocated the rate for each path in advance, which is not optimal.

In this section, we propose an algorithm to jointly optimize the source rate and the routing scheme for video unicasting over wireless ad hoc networks. Our algorithm uses dual decomposition to split the problem into multiple subproblems, and then solves each subproblem in parallel. The fully distributed nature of the proposed algorithm leads to two main advantages. First, it is very efficient; the optimal results converge quickly. This characteristic makes fast re-routing possible when the network topology changes due to mobility or channel failures. Secondly, it shares the computation load among all the nodes, thus saving the power consumption of each node.

5.2.1   Problem Formulation and Distributed Solution

5.2.1.1   Problem Formulation

We consider the joint optimization of the source rate and routing scheme for video unicasting over wireless ad hoc networks. The problem can be stated as follows.

Given a wireless ad hoc network $\mathbf{G} = (\mathbf{N}, \mathbf{L})$, and a unicast session from a source to a receiver, we want to minimize the expected video distortion d by optimally determining the source rate and the routing scheme, subject to the network flow constraint and the link capacity constraint. Mathematically, it is formulated as:

$$\begin{aligned}
\text{minimize}_{(s_r, \mathbf{x})} \quad & d = D_0 + \frac{\theta_0}{(s_r - \sum_{l \in \mathbf{L}} p_l x_l) + \phi_0} \\
\text{subject to} \quad & \sum_{l \in \mathbf{L}} a_{il} x_l = \eta_i, && \forall i \in \mathbf{N}, \\
& o \le x_l \le c_l, && \forall l \in \mathbf{L}, \\
& s_r \ge 0,
\end{aligned} \tag{5.10}$$

where s_r is the source rate, x_l is the flow rate at link l, p_l is the packet loss rate at link l, a_{il} represents the relationship between node i and link l, c_l is the link capacity at link l, and η_i is equal to s_r if node i is the source, or equal to $-s_r$ if node i is the receiver, or 0 for all other cases.

The optimization problem (5.10) can be converted to an equivalent Linear Programming (**LP**) problem as follows.

$$\begin{aligned}
\text{minimize}_{(s_r, \mathbf{x})} \quad & -s_r + \sum_{l \in \mathbf{L}} p_l x_l \\
\text{subject to} \quad & \text{the same constraints as in (5.10).}
\end{aligned} \tag{5.11}$$

In order to develop a distributed algorithm, we revise the objective function in problem (5.11) by adding a corresponding quadratic regularization term for each optimization variable. Then the optimization problem becomes:

$$\begin{aligned}
\text{minimize}_{(s_r, \mathbf{x})} \quad & -s_r + \sum_{l \in \mathbf{L}} x_l p_l + \delta s_r^2 + \delta \sum_{l \in \mathbf{L}} x_l^2 \\
\text{subject to} \quad & \text{the same constraints as in (5.10).}
\end{aligned} \tag{5.12}$$

where $\delta(\delta > 0)$ is the regularization factor.

The objective function in problem (5.12) is strictly convex. When δ is small enough, the solution for the optimization problem (5.12) is arbitrarily close to the solution for the optimization problem (5.10).

5.2.1.2 Distributed Solution

Since the optimization problem (5.12) has a coupled constraint, the dual decomposition technique [20] is appropriate to solve it. By Lagrangian relaxation, the optimization problem can be decoupled into several subproblems, which can be solved in parallel.

By introducing the dual variables $v_i, \forall i \in \mathbf{N}$ for each node, we have the Lagrangian of the primal problem (5.12) as follows.

$$L(s_r, \mathbf{x}, \mathbf{v}) = -s_r + \sum_{l \in \mathbf{L}} x_l p_l + \delta s_r^2 + \delta \sum_{l \in \mathbf{L}} x_l^2 + \sum_{i \in \mathbf{N}} v_i (\sum_{l \in \mathbf{L}} a_{il} x_l - \eta_i). \tag{5.13}$$

The Lagrange dual function $G(\mathbf{v})$ is the minimum value of the Lagrangian over primal variables s_r and $\mathbf{x}$.

$$\begin{aligned} G(\mathbf{v}) &= \min_{s_r, \mathbf{x}} \{ L(s_r, \mathbf{x}, \mathbf{v}) \} \\ &= \min_{s_r \geq 0} (-s_r + \delta s_r^2 - \sum_{i \in \mathbf{N}} v_i \eta_i) \\ &+ \sum_{l \in \mathbf{L}} \min_{0 \leq x_l \leq c_l} (\delta x_l^2 + x_l(p_l + \sum_{i \in \mathbf{N}} v_i a_{il})). \end{aligned} \tag{5.14}$$

The Lagrange dual problem is to maximize the Lagrange dual function, that is:

$$\text{maximize}_{(\mathbf{v})} \quad G(\mathbf{v}). \tag{5.15}$$

In the primal problem, the objective function is strictly convex and the constraints are linear. Therefore, Slater's condition for strong duality holds. The optimal duality gap is zero [19]. The primal variables s_r and $\mathbf{x}$ converge to the optimal solution of the primal problem (5.12) when a Lagrange dual problem (5.15) converges. At the kth iteration, the optimal primal variables s_r and $\mathbf{x}$ can be computed in parallel by solving the following problems.

$$s_r^{(k)} = \arg \min_{s_r \geq 0} \{ -s_r + \delta s_r^2 - \sum_{i \in \mathbf{N}} v_i^{(k)} \eta_i \}, \tag{5.16}$$

$$x_l^{(k)} = \arg \min_{0 \leq x_l \leq c_l} \{ \delta x_l^2 + x_l(p_l + \sum_{i \in \mathbf{N}} v_i^{(k)} a_{il}) \}, \quad \forall l \in \mathbf{L}. \tag{5.17}$$

We use a subgradient method [21] to solve a dual problem (5.15). The subgradient method is very efficient due to little requirements of memory usage and amenability for distributed implementation [21]. A subgradient of the negative dual function $-G(v)$ at $v_i^{(k)}$ is given by

$$g_i^{(k)} = \eta_i^{(k)} - \sum_{l \in \mathbf{L}} a_{il} x_l^{(k)}, \quad \forall i \in \mathbf{N}. \tag{5.18}$$

Dual variable $v_i^{(k+1)}$ at the $(k+1)^{th}$ iteration is updated by

$$v_i^{(k+1)} = v_i^{(k)} - \zeta^{(k)} g_i^{(k)}, \quad \forall i \in \mathbf{N}. \tag{5.19}$$

where $\zeta^{(k)} > 0$ is the step-size at the kth iteration. For a diminishing step-size: $\zeta^{(k)} = (1 + \rho)/(k + \rho)$, where $\rho > 0$, the algorithm is guaranteed to converge to the optimal value [21].

The above algorithm performs in a fully distributed way. First, the source node computes the optimal source rate using only its local dual variable. Second, each node computes the optimal flow rate of its outgoing links, using the packet loss rates of its outgoing links, the dual variables of itself and its neighboring nodes. If there is a topology change due to mobility or channel failures, our algorithm updates the dual variables, which then triggers the updates of the source rate and the link rates. After a quick adjustment, our algorithm can obtain the optimal solution corresponding to the new topology.

5.2.2   Simulation results

We simulate a wireless ad hoc network by randomly placing 10 nodes in a square region of 500-by-500m. We set the coverage threshold to be 250m. Two nodes can connect to each other if their distance is less than the coverage threshold. The available bandwidth of each link is randomly generated based on a Gaussian distribution with a mean of $800 \log_2(1 + (100/d_l)^2)$ Kbps where d_l is the distance in meters between the transmitter and receiver, and a variance $\sigma^2 = 20$. For each link, the transition probability from a GOOD state to the next GOOD state is uniformly distributed between $[0.85, 0.95]$, and the transition probability from a BAD state to the next BAD state is uniformly distributed between $[0.05, 0.15]$. We encode Foreman QCIF sequence into 8 layers using SNR-scalable extension of H.264/AVC [14]. The source bits are packetized into 8 descriptions, each having an average bit rate of 0.23 Mbps. Network coding is performed at the source node and the intermediate relay nodes. In the optimization, we set δ to 0.01, and ρ to 0.35. The convergence threshold for the dual function is set to 10^{-5}.

Figure 5.2 shows the iteration of the primal variables. After 260 iterations, the optimal source rate converges to 1.130 Mbps, as depicted in Figure 5.2(a). The iterations of all link rates are shown in Figure 5.2(b).

We compare the proposed multi-path routing scheme to the other two multi-path routing schemes: 1) congestion-minimized routing [9], where the link flow is computed by minimizing $\Sigma_{l \in \mathbf{L}}(x_l/(c_l - x_l))$; and 2) double-disjoint-path routing [24], where we find two disjoint paths by maximizing the end-to-end available bandwidth. In congestion-minimized routing, the maximal source rate that guarantees the convergence is 1.0 Mbps, the average Peak Signal-to-Noise Ratio (PSNR) of Foreman 300-frame QCIF video reconstructed at the receiver is 31.51 dB. Double-disjoint-path routing achieves a maximal source rate of 0.59 Mbps, and an average PSNR of 26.38 dB. The

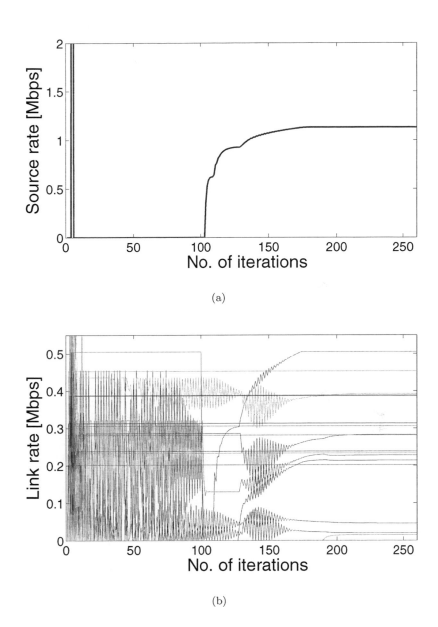

(a)

(b)

FIGURE 5.2
Iterations of the primal variables for video unicasting over a wireless ad hoc network: (a) source rate, and (b) link rates

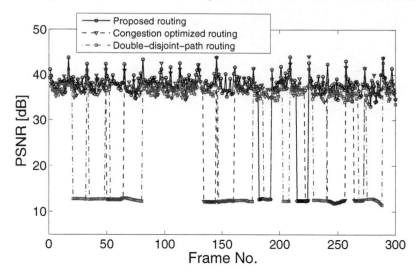

FIGURE 5.3
PSNR comparison among the proposed scheme, congestion-minimized routing scheme, and double-disjoint-path routing scheme, for video unicasting over a wireless ad hoc network

proposed routing outperforms the other two routing schemes with an average PSNR of 36.28 dB. The reason is that the proposed scheme fully utilizes the link bandwidth in a loss-aware manner. The PSNR of the 300 frames for these three routing schemes is plotted in Figure 5.3.

5.3   Optimized Video Multicasting over FDMA Wireless Ad Hoc Networks

In video multicast streaming, a video is transmitted in real-time simultaneously to multiple receivers. Video multicasting has many applications. For example, a group of visitors in a museum would like to receive real-time video stream (e.g., information about the collections from the museum guides) with their PDAs. The museum, requiring no preexisting infrastructure, can provide this service to a specific group of visitors via video multicasting over a wireless ad hoc network. As a result, video multicasting over wireless ad hoc networks provides a flexible solution to multiple users with a reduced cost.

Multicast over ad hoc networks is bandwidth-efficient compared to multiple unicast sessions. However, there are challenges for video multicasting over wireless ad hoc networks. First of all, it is difficult to find the optimal routing scheme in wireless ad hoc networks. Multi-path routing can provide a higher throughput to the receivers [25, 26]. However, how to optimally allocate the traffic to each path is a nontrivial task. Secondly, source rate allocation is another important problem for video multicasting over wireless ad hoc networks. If the source rate exceeds the network capacity, congestion will occur. On the other hand, if the source rate is too small, some users may not receive the video at maximum quality levels.

Video multicasting over wireless ad hoc networks has been studied in recent years. Most of the work uses either a single tree or multiple trees to perform video multicasting. However, the optimal tree-construction is NP-hard [25], and requires centralized computation. This is not suitable for a wireless ad hoc network, where each node has a limited computation capacity, and a central powerful node does not exist. The source rate allocation is under-investigated for video multicast in wireless ad hoc network. Most people use only two paths to represent multi-path routing, and allocate a source rate to each path in advance. Optimal traffic assignment into more paths (e.g., > 2) is expected to improve the performance.

In this section, we study the video multicasting over FDMA wireless ad hoc networks. Specifically, we propose a distributed algorithm to maximize the aggregate throughput by jointly optimizing the source rate allocation and the routing scheme. The proposed algorithm is fully distributed, making it extremely suitable for wireless ad hoc networks.

5.3.1 Problem Formulation and Distributed Solution

5.3.1.1 Problem Formulation

We adopt a Gaussian broadcast channel with FDMA as the communication model. Each Link is assigned with a disjoint frequency band. Based on Shannon theory, the capacity of link l can be formulated as follows [21].

$$c_l = W_l \log_2(1 + \frac{G_l P_l}{n_l^s W_l}), \quad \forall l \in \mathbf{L}, \tag{5.20}$$

where W_l is the bandwidth assigned to link l, P_l is the transmit power at the transmitter of link l, n_l^s is the power spectral density of additive white Gaussian noise at the receiver of link l, and G_l represents the path gain from the transmitter of link l to the receiver of link l.

As described in [2], the combination of the prioritized coding scheme and network coding eliminates the decoding hierarchy and delivery redundancy. A receiver with a larger throughput can reconstruct the video at a higher quality since all the received packets are distinct. For video multicasting over wireless ad hoc networks, our objective can be placed on the maximization of the aggregate throughput received at all of the receivers. We state the problem as follows.

Given a FDMA wireless ad hoc network $\mathbf{G} = (\mathbf{N}, \mathbf{L})$, and a multicast session from a source to H receivers, we want to maximize the aggregate throughput by optimally determining the source rate and the routing scheme, subject to the network flow constraint and the link capacity constraint. Mathematically, it is formulated as:

$$
\begin{aligned}
\text{maximize}_{(s,x)} \quad & \sum_{h \in \mathbf{V}}(s_h - p_h^E) = \sum_{h \in \mathbf{V}}(s_h - \sum_{l \in \mathbf{L}} x_{hl} p_l) \\
\text{subject to} \quad & \sum_{l \in \mathbf{L}} a_{il} x_{hl} = \eta_{hi}, && \forall h \in \mathbf{V}, \forall i \in \mathbf{N}, \\
& o \le x_{hl} \le c_l, && \forall h \in \mathbf{V}, \forall l \in \mathbf{L}, \\
& s_h \ge 0, && \forall h \in \mathbf{V},
\end{aligned}
\tag{5.21}
$$

where s_h is the source rate for conceptual session h, x_{hl} is the flow rate at link l for conceptual session h, p_h^E is the end-to-end packet loss rate for conceptual session h, p_l is the packet loss rate at link l, a_{il} represents the relationship between node i and link l, c_l is the link capacity at link l, given by Equation (5.20), η_{hi} is equal to s_h if node i is the source, or equal to $-s_h$ if node i is the receiver, or 0 for all other cases.

The optimization problem (5.21) is an **LP** problem, which can be approximated to a strictly convex optimization as follows.

$$
\begin{aligned}
\text{minimize}_{(s,x)} \quad & \sum_{h \in \mathbf{V}}(-s_h + \sum_{l \in \mathbf{L}} x_{hl} p_l + \delta s_h^2 + \delta \sum_{l \in \mathbf{L}} x_{hl}^2) \\
\text{subject to} \quad & \text{the same constraints as in (5.21).}
\end{aligned}
\tag{5.22}
$$

where $\delta(\delta > 0)$ is the regularization factor. When δ is sufficiently small, the term $(\delta s_h^2 + \delta \sum_{l \in \mathbf{L}} x_{hl}^2)$ is close to 0, and the solution for problem (5.22) is arbitrarily close to the solution for the original problem (5.21).

5.3.1.2   Distributed Solution

Since the objective function is strictly convex and the constraints are linear, the problem (5.22) represents a strictly convex optimization problem. By using the Lagrangian duality properties, we can develop a distributed algorithm to solve problem (5.22).

By introducing dual variables $v_{hi}, h \in \mathbf{V}, i \in \mathbf{N}$, we have the Lagrangian of the primal problem (5.22) as below.

$$
\begin{aligned}
L(\mathbf{s},\mathbf{x},\mathbf{v}) \quad &= \sum_{h \in \mathbf{V}}(-s_h + \sum_{l \in \mathbf{L}} x_{hl} p_l + \delta s_h^2 + \delta \Sigma_{l \in \mathbf{L}} x_{hl}^2) + \\
& \quad \sum_{h \in \mathbf{V}} \sum_{i \in \mathbf{N}} v_{hi}(\sum_{l \in \mathbf{L}} a_{il} x_{hl} - \eta_{hi}) \\
&= \sum_{h \in \mathbf{V}}(-s_h + \delta s_h^2 - \sum_{i \in \mathbf{N}} v_{hi} \eta_{hi}) + \\
& \quad \sum_{h \in \mathbf{V}} \sum_{l \in \mathbf{L}}(\delta x_{hl}^2 + x_{hl}(p_l + \sum_{i \in \mathbf{N}} v_{hi} a_{il})),
\end{aligned}
\tag{5.23}
$$

where $\mathbf{v}$ is the dual variable matrix, $\mathbf{s}$ is the vector of the conceptual source rates, and $\mathbf{x}$ is the matrix of the conceptual link rates.

The Lagrange dual function $G(\mathbf{v})$ is the minimum value of the Lagrangian over primal variables $\mathbf{s}$ and $\mathbf{x}$.

$$
\begin{aligned}
G(\mathbf{v}) \quad &= \min_{\mathbf{s},\mathbf{x}}\{L(\mathbf{s},\mathbf{x},\mathbf{v})\} \\
&= \sum_{h \in \mathbf{V}} \min_{s_h \ge 0}(-s_h + \delta s_h^2 - \sum_{i \in \mathbf{N}} v_{hi} \eta_{hi}) + \\
& \quad \sum_{h \in \mathbf{V}} \sum_{l \in \mathbf{L}} \min_{0 \le x_{hl} \le c_l}(\delta x_{hl}^2 + x_{hl}(p_l + \sum_{i \in \mathbf{N}} v_{hi} a_{il})).
\end{aligned}
\tag{5.24}
$$

The dual function $G(\mathbf{v})$ can be evaluated separately via the conceptual source rate s_h and the link rate x_{hl}. The Lagrange dual problem associated with the primal problem (5.22) is given by

$$\text{maximize}_{(\mathbf{v})} \quad G(\mathbf{v}) = G_{source}(\mathbf{v}) + G_{routing}(\mathbf{v}). \tag{5.25}$$

Since the dual function $G(\mathbf{v})$ is always concave, the Lagrange dual problem is therefore a convex optimization problem [19]. In the optimization problem (5.22), Slater's condition for strong duality holds [19]. The primal variables $\mathbf{s}$ and $\mathbf{x}$ converge to the optimal solution to the primal problem (5.22) when the dual variables $\mathbf{v}$ converge to the optimal solution to the dual problem (5.25).

At the kth iteration, the optimal primal variables can be obtained from dual variables.

$$s_h^{(k)} = \arg \min_{s_h \geq 0} \{-s_h + \delta s_h^2 - \sum_{i \in \mathbf{N}} v_{hi}^{(k)} \eta_{hi}\}, \quad \forall h \in \mathbf{V}. \tag{5.26}$$

$$x_{hl}^{(k)} = \arg \min_{0 \leq x_{hl} \leq c_l} \{\delta x_{hl}^2 + x_{hl}(p_l + \sum_{i \in \mathbf{N}} v_{hi}^{(k)} a_{il})\}, \quad \forall h \in \mathbf{V}, \forall l \in \mathbf{L}. \tag{5.27}$$

We use the subgradient method [21] to solve the dual problem (5.25). Since the dual function is continuously differentiable due to the strictly convexity of the primal objective function, we can find a subgradient from the gradient. The subgradient of the negative dual function $-G(\mathbf{v})$ at v_{hi} is given by

$$g_{hi}^{(k)} = \eta_{hi}^{(k)} - \sum_{l \in \mathbf{L}} a_{il} x_{hl}^{(k)}, \quad \forall h \in \mathbf{V}, \forall i \in \mathbf{N}. \tag{5.28}$$

We update the dual variables by

$$v_{hi}^{(k+1)} = v_{hi}^{(k)} - \theta^{(k)} g_{hi}^{(k)}, \quad \forall h \in \mathbf{V}, \forall i \in \mathbf{N}. \tag{5.29}$$

where $\theta^{(k)} > 0$ is the step-size at k^{th} iteration. One simple convergence condition requires that the step-size sequence satisfies the non-summable diminishing rule [20]:

$$\lim_{k \to \infty} \theta^{(k)} = 0, \quad \sum_{k=1}^{\infty} \theta^{(k)} = \infty. \tag{5.30}$$

The above algorithm is fully distributed in the following senses. First, the source node computes each conceptual source rate using its dual variable and the dual variable of the receiver. Second, each node computes the conceptual link rates of its outgoing links, using the packet loss rates of its outgoing links, the local dual variable, and the dual variables of its neighboring nodes. The proposed distributed algorithm only requires information exchange in the neighborhood, thus greatly reducing the overhead.

Receiver ID	Optimized routing scheme	Double-tree routing scheme
Node 6	31.68 dB	29.94 dB
Node 7	32.39 dB	29.83 dB
Node 10	33.73 dB	28.87 dB
Node 11	33.52 dB	29.32 dB

TABLE 5.1
Comparison of average PSNR between the optimized routing scheme and the double-tree routing scheme for video multicasting over a FDMA wireless ad hoc network

5.3.2   Simulation Results

We generate a wireless ad hoc network by placing 15 nodes at random locations in a square region of 400m-by-400m. Two nodes are able to communicate if their distance is smaller than the coverage threshold 150m. Node 1 is chosen as the source node. Four nodes (node 6, 7, 10 and 11) are chosen as the receivers. For every link, the transition probability from a GOOD state to the next GOOD state is uniformly distributed in [0.90, 0.95], and the transition probability from a BAD state to the next BAD state is uniformly distributed in [0.05, 0.10]. We encode Foreman QCIF sequence into 8 layers using SNR-scalable extension of H.264/AVC [14]. Each GOP consists of 16 frames. The source bits are packetized into 8 descriptions, each having an average bit rate of 67.0 Kbps. No error-concealment is used for the lost frames. In the communication model, the bandwidth allocated for each link is 15 KHz. The transmit power at each link is fixed at 0.70 W. The power spectral density n_l^s of additive white Gaussian noise at each receiver is uniformly distributed in [0.05, 0.10] W/Hz. The path gain G_l for link l is given by $G_l = 0.01/d_l^2$, where d_l is the distance in meters between the transmitter and the receiver at link l. In the optimization, regularization factor δ is set to 0.05. The step-size at the kth iteration is $\theta^{(k)} = 0.25/\sqrt{k}$. The convergence threshold for the dual function is set to 10^{-5}.

With the predefined convergence threshold, all the primal variables (the conceptual source rates and link rates) converge within 155 iterations. The fast convergence speed makes fast re-routing possible when the network topology changes due to mobility or channel failures. The iteration of the conceptual source rates is shown in Figure 5.4(a). The maximum among all the optimal conceptual source rates is the optimal multicast source rate, which is 0.565 Mbps. There are totally 80 links in the ad hoc network. The multicast link rate at each link is the maximum among four conceptual unicast link rates. We can see from Figure 5.4(b) that all the link rate variables converge within 155 iterations.

We compare the proposed optimized routing scheme to the double-tree routing scheme. In the double-tree routing scheme, we construct the double trees as follows. All the nodes except the source are classified into two cat-

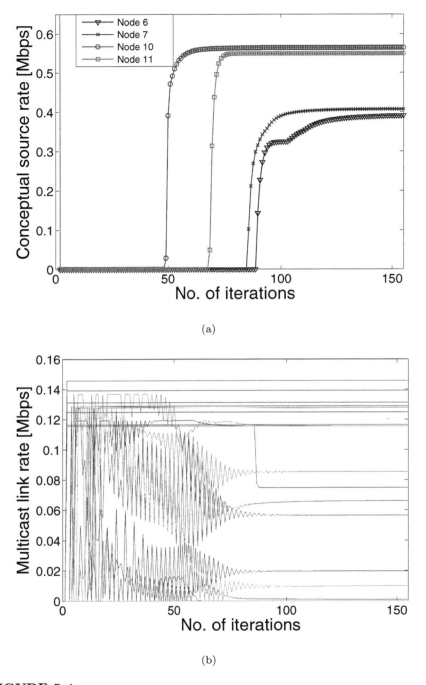

(a)

(b)

FIGURE 5.4
Iterations of primal variables for video multicasting over a FDMA wireless ad
hoc network: (a) conceptual source rates, and (b) multicast link rates

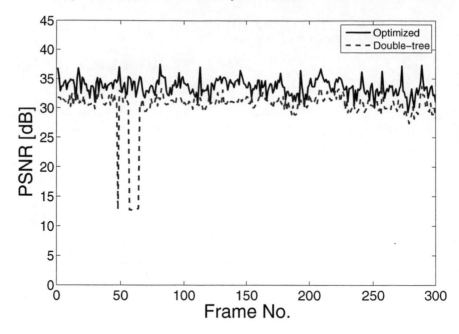

FIGURE 5.5
Frame PSNR comparison at node 11 for video multicasting over a FDMA
wireless ad hoc network

egories: group 0 and group 1. Within each group, we construct a single tree
from the source to the receivers by using the link throughput metric $c_l(1 - p_l)$.
The source rate for each conceptual session in the double-tree scheme is equal
to the end-to-end available bandwidth. In order for fair comparison, the bandwidth consumption and power consumption at each node are kept the same in
both schemes. Table 6.1 shows the comparison of average PSNR between the
two schemes. The optimized routing scheme improves the PSNR by 3.34 dB in
average compared to the double-tree scheme. The frame PSNR comparison of
the reconstructed Foreman QCIF sequence at node 11 is illustrated in Figure
5.5. The optimized routing scheme yields higher PSNR values since it yields
an improved end-to-end throughput.

5.4 Distributed Cross-Layer Optimization for Video Multicasting over CDMA Wireless Ad Hoc Networks

Video streaming over wireless ad hoc networks has been studied extensively in the recent decade. Layered video coding combined with a selective Automatic Repeat Request (ARQ) scheme has been proposed in [16], in which the base-layer and the enhancement-layer packets are transmitted over different paths, and only the base-layer packets are retransmitted. However, if the Round-Trip-Time (RTT) is large, ARQ is not applicable to real-time streaming. Multiple Description Coding (MDC) is another coding technique widely used in multi-path video transport over wireless ad hoc networks [25][6]. The receiver can reconstruct the video at a low quality when receiving fewer descriptions, and reconstruct the video at a high quality when more descriptions are received. Multi-path transport with MDC is resilient to network outages in the wireless ad hoc networks. MDC multicast in wireless ad hoc networks has been studied in [25], where the authors construct multiple multicast trees, each delivering one description.

Multicast routing has been an active research area for many years [17]. Most of these problems belong to the class of minimum or constrained minimum Steiner tree problems which are well known to be NP-complete. Most of the algorithms aim to find a single tree using network layer performance metrics, such as delay, loss, or throughput. The single-tree routing cannot make good use of the available link bandwidth, thus cannot provide a good performance for video streaming. Recently, multiple tree routing algorithms are proposed to explore the path diversity for each receiver. Two typical multiple-tree video multicast schemes in wireless ad hoc networks are introduced in [25] and [26], respectively. In [25], two multicast trees are constructed to deliver two descriptions. Each description is layered, encoded to meet the heterogeneous capacity of the receivers. The authors optimize the expected video quality using a genetic algorithm when constructing multiple trees. However, the computation is essentially centralized and computationally intensive. In [26], the authors propose a multiple tree construction protocol that builds two nearly disjoint trees simultaneously in a distributed manner. However, the trees are built based on the network layer metrics, and the application performance has not yet been optimized.

Optimization techniques are widely used in the video streaming applications. Some of the optimizations require centralized information, making them unsuitable for large network optimization. Dual decomposition has been applied in network optimizations. Kelly et al. have investigated two classes of distributed rate control algorithms corresponding to the primal and dual decomposition of the optimization [18]. Chiang et al. use convex optimization in the framework of network utility maximization [19]. Cross-layer optimization in wireless ad hoc networks has been presented in [20][21][22]. However, most of the work assumes that the utility function is strictly convex. The optimization for multiple unicast video streams in wireless ad hoc networks is studied in [23].

In this section, we present a distributed cross-layer optimization for video multicasting over CDMA wireless ad hoc networks. Different from Section 5.3, we take into account the power allocation in physical layers in this section. Specifically, we propose a distributed algorithm to jointly optimize the source rate allocation, the routing scheme and the power allocation. We use hierarchical dual decompositions to separate the optimization problem into multiple subproblems, which are then solved in parallel. The distributed nature of the proposed algorithm leads to a fast convergence of the optimal results.

5.4.1   Problem Formulation

For video multicasting over a CDMA wireless ad hoc network, we consider three important issues: the source rate allocation, the routing schemes and the power allocation. The three issues are inter-related. The link rate needs to be no larger than the link capacity, which is dependent on the power allocation. In addition, the link rate allocation is dependent on the source rate; a different routing scheme accommodates a different maximum source rate. Therefore, we consider the three issues in a jointly optimized way.

The combination of the prioritized coding scheme and network coding eliminates the decoding hierarchy and delivery redundancy. Therefore, a larger throughput at a receiver can lead to a smaller distortion. We set the objective as maximizing the aggregate throughput, and state the problem of Joint Optimization of Source rate, Routing and Power (**JOSRP**) as follows.

Given a CDMA wireless ad hoc network $\mathbf{G} = (\mathbf{N}, \mathbf{L})$, and a multicast session from a source to H receivers, we want to maximize the aggregate throughput by optimally determining the source rate, the routing scheme and the power allocation.

Mathematically, the problem of **JOSRP** is formulated as:

$$
\begin{aligned}
\text{maximize}_{(\mathbf{s},\mathbf{x},\mathbf{y},\mathbf{P})} \quad & \sum_{h\in\mathbf{V}}\left(s_h - \sum_{l\in\mathbf{L}} p_l x_{hl}\right) \\
\text{subject to} \quad & \sum_{l\in\mathbf{L}} a_{il} x_{hl} = \eta_{hi}, \quad \forall h \in \mathbf{V}, \forall i \in \mathbf{N}, \\
& x_{hl} \le y_l, \quad \forall h \in \mathbf{V}, \forall l \in \mathbf{L}, \\
& y_l \le c_l, \quad \forall l \in \mathbf{L}, \\
& c_l = f(\mathbf{P}), \quad \forall l \in \mathbf{L}, \\
& x_{hl} \ge 0, \quad \forall h \in \mathbf{V}, \forall l \in \mathbf{L}, \\
& s_h \ge 0, \quad \forall h \in \mathbf{V}, \\
& 0 \le P_l \le P_m, \quad \forall l \in \mathbf{L},
\end{aligned}
\tag{5.31}
$$

where s_h is the source rate for conceptual session h, x_{hl} is the flow rate at link l for conceptual session h, y_l is the multicast rate at link l, p_l is the PLR at link l, a_{il} represents the relationship between node i and link l, c_l is the link capacity at link l, P_m is the upper bound of transmit power, η_{hi} is equal to s_h

if node i is the source, or equal to $-s_h$ if node i is the receiver for conceptual session h, or 0 otherwise.

In the constraints of the **JOSRP** problem, $\sum_{l \in \mathbf{L}} a_{il} x_{hl} = \eta_{hi}$ represents the flow conservation, $x_{hl} \leq y_l$ shows that the multicast rate at link l is no less than any of the conceptual unicast rates going through this link, $y_l \leq c_l$ shows that the multicast rate is no larger than the link capacity, which is determined by the power vector $\mathbf{P}$. We solve the **JOSRP** problem in an iterative way. At each iteration, a **JOSRP** problem is reconstructed based on the current PLRs, which are computed using the transmit powers obtained at the previous iteration.

5.4.2 Distributed Algorithm

In this section, we present a distributed solution to problem (5.31). Our solution framework consists of two layers of decompositions. First, we decompose the main problem (5.31) into the network flow subproblem and the power allocation subproblem using dual decomposition. For the network flow subproblem, we perform second-layer decomposition and solve it in a distributed way. For the power allocation problem, we employ a game theoretic approach to solve it.

5.4.2.1 First-Layer Decomposition

By introducing the dual variable $\lambda_l (\forall l \in \mathbf{L})$ for the capacity constraint $y_l \leq c_l, \forall l \in \mathbf{L}$, we have the Lagrangian [19] corresponding to the primal problem (5.31) as follows.

$$
\begin{aligned}
&L(\mathbf{s}, \mathbf{x}, \mathbf{y}, \mathbf{P}, \lambda) \\
&= \sum_{h \in \mathbf{V}} (-s_h + \sum_{l \in \mathbf{L}} p_l x_{hl}) + \sum_{l \in \mathbf{L}} \lambda_l (y_l - c_l) \\
&= \sum_{h \in \mathbf{V}} (-s_h + \sum_{l \in \mathbf{L}} p_l x_{hl}) + \sum_{l \in \mathbf{L}} \lambda_l y_l - \sum_{l \in \mathbf{L}} \lambda_l c_l.
\end{aligned}
\tag{5.32}
$$

The Lagrange dual function $G(\lambda)$ is the minimum value of the Lagrangian over the primal variables $(\mathbf{s}, \mathbf{x}, \mathbf{y}, \mathbf{P})$.

$$
G(\lambda) = \min_{(\mathbf{s}, \mathbf{x}, \mathbf{y}, \mathbf{P})} L(\mathbf{s}, \mathbf{x}, \mathbf{y}, \mathbf{P}, \lambda).
\tag{5.33}
$$

The Lagrange dual function $G(\lambda)$ can be evaluated separately in the network flow variables $(\mathbf{s}, \mathbf{x}, \mathbf{y})$ and the power variables $\mathbf{P}$, which is expressed by

$$
G(\lambda) = G_{network}(\lambda) + G_{power}(\lambda).
\tag{5.34}
$$

In Equation (5.34), $G_{network}(\lambda)$ is given by

$$
\begin{aligned}
&G_{network}(\lambda) \\
&= \min_{(\mathbf{s}, \mathbf{x}, \mathbf{y})} \quad \{ \sum_{h \in \mathbf{V}} (-s_h + \sum_{l \in \mathbf{L}} p_l x_{hl}) + \sum_{l \in \mathbf{L}} \lambda_l y_l, \\
&\qquad\qquad | \text{ network constraint set} \}
\end{aligned}
\tag{5.35}
$$

where the network constraint set consists of 1) $\sum_{l\in\mathbf{L}} a_{il}x_{hl} = \eta_{hi}, \forall h \in \mathbf{V}, \forall i \in \mathbf{N}$; 2) $x_{hl} \le y_l, \forall h \in \mathbf{V}, \forall l \in \mathbf{L}$; 3) $x_{hl} \ge 0, \forall h \in \mathbf{V}, \forall l \in \mathbf{L}$; and 4) $s_h \ge 0, \forall h \in \mathbf{V}$.

In Equation (5.34), $G_{power}(\lambda)$ is given by

$$G_{power}(\lambda) = \min_{(\mathbf{P})}\{\sum_{l\in\mathbf{L}}(-\lambda_l c_l),|\ 0 \le P_l \le P_m, \forall l \in \mathbf{L}\}. \tag{5.36}$$

The Lagrange dual problem associated with the primal problem (5.31) is to maximize the Lagrange dual function, that is:

$$\begin{aligned} &\text{maximize}_{(\lambda)} \quad G(\lambda) = G_{network}(\lambda) + G_{power}(\lambda) \\ &\text{subject to} \quad \lambda_l \ge 0, \ \forall l \in \mathbf{L}. \end{aligned} \tag{5.37}$$

Subgradient method [21] is used to solve the Lagrange dual problem (5.37). Dual variable λ_l is given an initial value $\lambda_l^{(0)}$, and is iteratively updated by

$$\lambda_l^{(t+1)} = \max\{0, \lambda_l^{(t)} - \theta^{(t)}(c_l^{(t)} - y_l^{(t)})\}, \quad \forall l \in \mathbf{L}, \tag{5.38}$$

where $\theta^{(t)}$ is the step-size, $c_l^{(t)}$ is the link capacity at link l, and $y_l^{(t)}$ is the multicast link rate at link l at the tth iteration.

5.4.2.2 Second-Layer Decomposition

At the tth iteration, the first-layer dual variables are denoted by a vector $\lambda^{(t)}$. Given a $\lambda^{(t)}$, we can obtain the corresponding primal variables $(\mathbf{s}, \mathbf{x}, \mathbf{y}, \mathbf{P})$ respectively. Since the Lagrange dual function $G(\lambda)$ is separated into $G_{network}(\lambda)$ and $G_{power}(\lambda)$, we can find the network flow variables $(\mathbf{s}, \mathbf{x}, \mathbf{y})$ and the power variables $\mathbf{P}$ separately.

In order to find the network flow variables $(\mathbf{s}, \mathbf{x}, \mathbf{y})$, we need to solve the network flow subproblem as follows.

$$\begin{aligned} &\text{minimize}_{(\mathbf{s},\mathbf{x},\mathbf{y})} \quad \sum_{h\in\mathbf{V}}(-s_h + \sum_{l\in\mathbf{L}} p_l x_{hl}) + \sum_{l\in\mathbf{L}} \lambda_l^{(t)} y_l, \\ &\text{subject to} \quad \sum_{l\in\mathbf{L}} a_{il}x_{hl} = \eta_{hi}, \quad \forall h \in \mathbf{V}, \forall i \in \mathbf{N}, \\ &\qquad\qquad x_{hl} \le y_l, \quad \forall h \in \mathbf{V}, \forall l \in \mathbf{L}, \\ &\qquad\qquad x_{hl} \ge 0, \quad \forall h \in \mathbf{V}, \forall l \in \mathbf{L}, \\ &\qquad\qquad s_h \ge 0, \quad \forall h \in \mathbf{V}. \end{aligned} \tag{5.39}$$

By adding a quadratic regularization term for each optimization variable, we approximate the **LP** problem in (5.39) into a strictly convex optimization problem as follows.

$$\begin{aligned} &\text{minimize}_{(\mathbf{s},\mathbf{x},\mathbf{y})} \quad \sum_{h\in\mathbf{V}}(-s_h + \sum_{l\in\mathbf{L}} p_l x_{hl} + \delta s_h^2 + \\ &\qquad\qquad \delta\sum_{l\in\mathbf{L}} x_{hl}^2) + \sum_{l\in\mathbf{L}} \lambda_l^{(t)} y_l + \delta\sum_{l\in\mathbf{L}} y_l^2 \\ &\text{subject to} \quad \text{the same constraints as in (5.39).} \end{aligned} \tag{5.40}$$

We can make the solution for problem (5.40) arbitrarily close to the solution for problem (5.39) by choosing a sufficiently small $\delta(\delta > 0)$.

Similarly, dual decomposition is used in the second-layer decomposition. We introduce second-layer dual variables $u_{hl}(\forall h \in \mathbf{V}, \forall l \in \mathbf{L})$ and $v_{hi}(\forall h \in \mathbf{V}, \forall i \in \mathbf{N})$, and then formulate the Lagrangian corresponding to the primal problem (5.40) as below.

$$
\begin{aligned}
& L(\mathbf{s}, \mathbf{x}, \mathbf{y}, \mathbf{u}, \mathbf{v}) \\
&= \sum_{h \in \mathbf{V}} (-s_h + \sum_{l \in \mathbf{L}} p_l x_{hl} + \delta s_h^2 + \delta \sum_{l \in \mathbf{L}} x_{hl}^2) + \\
& \quad \sum_{l \in \mathbf{L}} \lambda_l^{(t)} y_l + \delta \sum_{l \in \mathbf{L}} y_l^2 + \\
& \quad \sum_{h \in \mathbf{V}} \sum_{i \in \mathbf{N}} v_{hi} (\sum_{l \in \mathbf{L}} a_{il} x_{hl} - \eta_{hi}) + \\
& \quad \sum_{h \in \mathbf{V}} \sum_{l \in \mathbf{L}} u_{hl} (x_{hl} - y_l).
\end{aligned}
\tag{5.41}
$$

The Lagrange dual function $D(\mathbf{u}, \mathbf{v})$ is formulated as:

$$
\begin{aligned}
& D(\mathbf{u}, \mathbf{v}) = \min_{(\mathbf{s}, \mathbf{x}, \mathbf{y})} L(\mathbf{s}, \mathbf{x}, \mathbf{y}, \mathbf{u}, \mathbf{v}) \\
&= \min_{(\mathbf{s})} \sum_{h \in \mathbf{V}} (-s_h + \delta s_h^2 - \sum_{i \in \mathbf{N}} v_{hi} \eta_{hi}) + \\
& \quad \min_{(\mathbf{x})} \sum_{h \in \mathbf{V}} \sum_{l \in \mathbf{L}} (\delta x_{hl}^2 + x_{hl}(p_l + u_{hl} + \sum_{i \in \mathbf{N}} v_{hi} a_{il})) + \\
& \quad \min_{(\mathbf{y})} \sum_{l \in \mathbf{L}} (\delta y_l^2 + y_l(\lambda_l^{(t)} - \sum_{h \in \mathbf{V}} u_{hl})).
\end{aligned}
\tag{5.42}
$$

The Lagrange dual problem associated with the primal problem (5.40) is given by

$$
\begin{aligned}
& \text{maximize}_{(\mathbf{u}, \mathbf{v})} \quad D(\mathbf{u}, \mathbf{v}) \\
& \text{subject to} \quad u_{hl} \geq 0, \ \forall h \in \mathbf{V}, \forall l \in \mathbf{L}.
\end{aligned}
\tag{5.43}
$$

Since the dual function is always concave, we can solve the dual problem with subgradient method [21]. The dual variables $u_{hl}^{(k+1)}$ and $v_{hi}^{(k+1)}$ at the $(k+1)^{th}$ iteration are updated respectively by

$$
u_{hl}^{(k+1)} = \max\{0, u_{hl}^{(k)} - \theta^{(k)}(y_l^{(k)} - x_{hl}^{(k)})\}, \quad \forall h \in \mathbf{V}, \forall l \in \mathbf{L},
\tag{5.44}
$$

$$
v_{hi}^{(k+1)} = v_{hi}^{(k)} - \theta^{(k)}(\eta_{hi}^{(k)} - \sum_{l \in \mathbf{L}} a_{il} x_{hl}^{(k)}), \quad \forall h \in \mathbf{V}, \forall i \in \mathbf{N},
\tag{5.45}
$$

where $\theta^{(k)} > 0$ is the step-size at the kth iteration. The algorithm is guaranteed to converge to the optimal value for a step-size sequence satisfying the non-summable diminishing rule [20]:

$$
\lim_{k \to \infty} \theta^{(k)} = 0, \quad \sum_{k=1}^{\infty} \theta^{(k)} = \infty.
\tag{5.46}
$$

The step-size we use in our algorithm is $\theta^{(k)} = \omega/\sqrt{k}$, where $\omega > 0$.

At the kth iteration, the optimal primal variables $(\mathbf{s}, \mathbf{x}, \mathbf{y})$ can be computed in parallel by solving the following problems separately.

$$
s_h^{(k)} = \arg \min_{(s_h \geq 0)} (-s_h + \delta s_h^2 - \sum_{i \in \mathbf{N}} v_{hi}^{(k)} \eta_{hi}), \quad \forall h \in \mathbf{V}.
\tag{5.47}
$$

$$
x_{hl}^{(k)} = \arg \min_{(x_{hl} \geq 0)} (\delta x_{hl}^2 + x_{hl}(p_l + u_{hl}^{(k)} + \sum_{i \in \mathbf{N}} v_{hi}^{(k)} a_{il})) \\
\forall h \in \mathbf{V}, \forall l \in \mathbf{L}.
\tag{5.48}
$$

$$y_l^{(k)} = \arg \min_{(y_l \geq 0)} (\delta y_l^2 + y_l(\lambda_l^{(t)} - \sum_{h \in \mathbf{V}} u_{hl}^{(k)})), \quad \forall l \in \mathbf{L}. \tag{5.49}$$

In order to obtain the power variables $\mathbf{P}$, we need to solve the power allocation subproblem as follows.

$$\begin{aligned}
\text{minimize}_{(\mathbf{P})} \quad & \sum_{l \in \mathbf{L}} (-\lambda_l^{(t)} c_l) \\
\text{subject to} \quad & c_l = B \log_2(1 + \frac{G_{ll} P_l}{\sum_{k \in \mathbf{L}, k \neq l} G_{lk} P_k + n_l}), \quad \forall l \in \mathbf{L}, \\
& 0 \leq P_l \leq P_m, \quad \forall l \in \mathbf{L}.
\end{aligned} \tag{5.50}$$

The problem in (5.50) is a non-convex optimization problem. We use a game theoretic approach [20] to solve it. Link l maximizes its payoff function as follows.

$$Q_l = \lambda_l^{(t)} B \log_2(1 + \frac{G_{ll} P_l}{\sum_{k \in \mathbf{L}, k \neq l} G_{lk} P_k + n_l}) - m_l P_l, \quad \forall l \in \mathbf{L}, \tag{5.51}$$

where m_l is the dual variable representing the interference effect to all the other links. A sensible choice for m_l is $\frac{\partial(\sum_{k \in \mathbf{L}, k \neq l} \lambda_k^{(t)} c_k)}{\partial P_l}$ [20]. The dual variable m_l can be explained as the rate at which the data rate at other links decreases with an additional amount of power at link l. The power allocation problem is solved iteratively as follows. Given the dual variable m_l^j at the jth iteration, we find $\mathbf{P}^{(j+1)}$ by setting the derivative of Q_l with respect to P_l to zero.

$$\widetilde{P}_l^{(i+1)} = \max\{0, \frac{\lambda_l^{(t)} B}{m_l^{(j)} \ln(2)} - \sum_{k \in \mathbf{L}, k \neq l} \frac{G_{lk}}{G_{ll}} \widetilde{P}_k^{(i)} - \frac{n_l}{G_{ll}}\}, \forall l \in \mathbf{L}. \tag{5.52}$$

We repeat the iteration in (5.52) until $\widetilde{\mathbf{P}}^{(i)}$ converges. Then we set $\mathbf{P}^{(j+1)} = \widetilde{\mathbf{P}}^{(i)}$. The dual variable $m_l^{(j+1)}$ at the $(j+1)^{th}$ iteration is updated by

$$\begin{aligned}
m_l^{(j+1)} & \\
= \sum_{k \in \mathbf{L}, k \neq l} & G_{kl} \psi_k \\
= \sum_{k \in \mathbf{L}, k \neq l} & (\frac{G_{kl} \lambda_k^{(t)} B}{\ln(2)})(\frac{SINR_k^{(j+1)}}{G_{kk} P_k^{(j+1)}})(\frac{SINR_k^{(j+1)}}{1 + SINR_k^{(j+1)}}), \forall l \in \mathbf{L},
\end{aligned} \tag{5.53}$$

where ψ_k represents a broadcast message, $SINR_k$ is the SINR at link k, which is given by $SINR_k = G_{kk} P_k / (\sum_{s \in \mathbf{L}, s \neq k} G_{ks} P_s + n_k)$.

In summary, the optimization algorithm is distributed in the following senses. In the first-layer decomposition, each node updates the dual variable λ_l for each of its outgoing links using the link capacity and the multicast link rate, as shown in Equation (5.38). Given the first-layer dual variables $\lambda^{(t)}$, the second-layer decomposition is performed as follows. First, the source node computes the source rate using Equation (5.47), each node computes the conceptual link rate x_{hl} using Equation (5.48), and the multicast link rate y_l using Equation (5.49) for each of its outgoing links, respectively. All the computations at each node use only the local information. In other words, each node collects the dual variables from its neighbors and computes the conceptual link rate and multicast link rate. Second, each node computes the broadcast message ψ_l for each of its outgoing links, and broadcasts ψ_l to the network. After collecting broadcast messages from all the other links, each node computes the dual variable m_l and updates the transmit power at each of its outgoing links. In the proposed algorithm, all the messages except broadcast message ψ_l are exchanged in the local neighborhood. The algorithm can converge quickly due to the decentralization, thus greatly reducing the overhead.

5.4.3 Simulation Results

We have carried out extensive simulations to investigate the performance of the proposed algorithm. We generate a wireless ad hoc network by placing 15 nodes at random locations in a square region of 400m by 400m. Two nodes are able to communicate to each other if their distance is smaller than a coverage threshold of 150m. Node 1 is chosen as the source node. Four receivers (node 6, 7, 10 and 11) are chosen at the edge of the network.

For every link, the transition probability from a GOOD state to the next GOOD state is uniformly distributed between [0.90, 0.95], and the transition probability from a BAD state to the next BAD state is uniformly distributed between [0.05, 0.10]. In the CDMA model, the maximum transmit power at each link is $P_m = 2.0W$, the baseband bandwidth for each link is 30 KHz. The gain for each link is computed by $G_{ij} = (1/220)(1/d_{ij})^4$ for $i \neq j$, and $G_{ii} = (1/d_{ij})^4$, where d_{ij} is the distance from the transmitter of link j to the receiver of link i, the factor of $(1/220)$ can be viewed as the spreading gain in a CDMA system. We encode a Foreman QCIF sequence into 8 layers using SNR-scalable extension of H.264/AVC [14]. Each GOP consists of 16 frames. The source bits are packetized into 8 descriptions in a prioritized way, each having an average bit rate of 66.7 Kbps. The source packets are encoded using random linear network coding before they are sent out by the source node. The intermediate nodes also perform network coding before they forward the received packets.

The convergence threshold for the dual function is set to 10^{-5}. Figure 5.6 illustrates the convergence of the proposed distributed algorithm. We can see that all the primal variables and dual variable converge within 497 iterations. Figure 5.6(a) shows the convergence of the 4 conceptual source rates. The multicast source rate is the maximum among the 4 conceptual source rates. There are 36 links in the ad hoc network. In order for clear presentation, we randomly choose 10 links and present the iterations of the multicast link rates and the transmit powers, which all converge within 497 iterations.

We compare the performance of the optimized scheme with the uniform-power scheme, in which each link is allocated with the same power, and the total power consumption is equal to that in the optimized scheme. Figure 5.7(a) shows the comparison of the end-to-end throughput for different receivers.

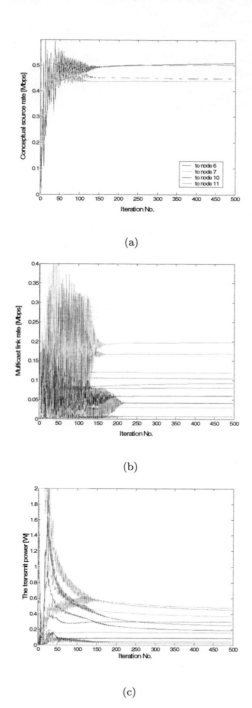

FIGURE 5.6
Iterations of (a) conceptual source rates, (b) multicast link rates, and (c) transmit powers, for video multicasting over a CDMA wireless ad hoc network

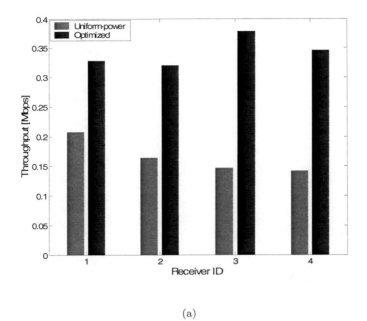

(a)

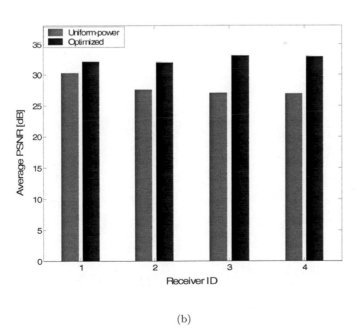

(b)

FIGURE 5.7
Comparison between the optimized scheme and the uniform-power scheme
for video multicasting over a CDMA wireless ad hoc network: (a) throughput
comparison, and (b) average PSNR comparison

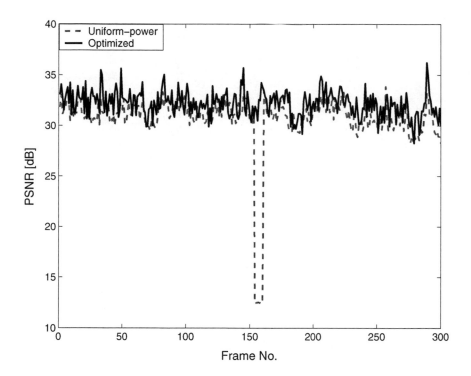

FIGURE 5.8

Frame PSNR comparison at node 6 between the optimized scheme and the uniform-power scheme for video multicasting over a CDMA wireless ad hoc network

(a)

(b)

FIGURE 5.9
Comparison of reconstructed frame 26 of Foreman QCIF sequence at node
6 between the optimized scheme and the uniform-power scheme for video
multicasting over a CDMA wireless ad hoc network: (a) the uniform-power
scheme (30.097 dB) and (b) the optimized scheme (33.706 dB)

The optimized scheme allocates the powers optimally, which produces an optimal allocation of link capacity. Therefore, the optimized scheme achieves a higher throughput compared to the uniform-power scheme. Since network coding eliminates the duplicate packets, a higher throughput leads to a higher average PSNR, which is depicted in Figure 5.7(b). We also present the frame PSNR comparison at node 6 between the two schemes in Figure 5.8. In the optimized scheme, the receivers can receive more packets for each frame, thus reconstructing the video at an overall higher PSNR. On the other hand, the receiver in the uniform-power scheme receives fewer packets for each frame due to a lower throughput, thus reconstructing the video at a worse quality. The subjective quality of reconstructed frame 26 of Foreman QCIF sequence is compared in Figure 5.9. The optimized scheme enable node 6 to reconstruct the frame at the PSNR of 33.706 dB, much higher than the PSNR of 30.097 dB, produced by the uniform-power scheme.

We also compare the optimized scheme with the tree-based routing schemes. In the optimized scheme, the link flow is obtained from the distributed optimization algorithm. In the single-tree based routing scheme, we construct the tree using the link throughput metric $c_l(1 - p_l)$. We choose the optimized single tree, where the summation of the end-to-end throughput from the source to all receivers is maximized. In the multiple-tree based routing scheme, we construct double trees as follows. All the nodes except the source are classified into two categories: group 0 and group 1. Within each group, we construct a single tree from the source to the receivers by using the link throughput metric $c_l(1 - p_l)$. In the above three transmission schemes, the total power consumption is kept the same for a fair comparison.

The comparison of the throughput and average PSNR among the three schemes are shown in Figure 5.10. The optimized scheme yields a much higher throughput, thus reconstructing the video at a higher average PSNR. In the double-tree scheme, all the receivers receive streams through two paths, thus having a higher end-to-end throughput compared to the single-tree scheme. Figure 5.11 compares the frame PSNR at node 10 between the optimized scheme and the double-tree scheme. We can see the frame PSNR in the optimized scheme is much higher than that in the double-tree scheme. In the double-tree scheme, the frame loss occurs frequently, and the drifting errors make the rest frames in that GOP undecodable. So we see many quality drops in the PSNR trace of the double-tree scheme. The subjective quality of frame 13 for these three schemes is compared in Figure 5.12. Optimized scheme produces the frame PSNR of 34.755 dB, much higher than 31.282 dB in the double-tree scheme, and 30.747 dB in the single-tree scheme. Therefore, we observe a much clearer picture in Figure 5.12(c), compared to both pictures in Figure 5.12(a) and Figure 5.12(b).

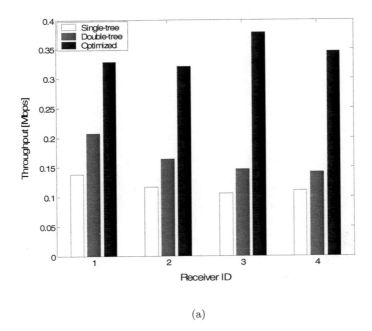

(a)

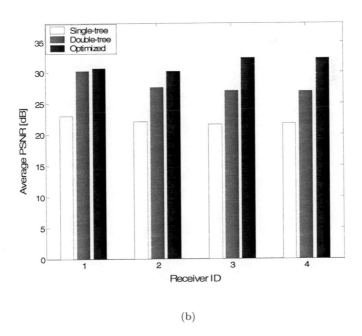

(b)

FIGURE 5.10
Comparison between the optimized scheme and the tree-based schemes for video multicasting over a CDMA wireless ad hoc network: (a) throughput comparison, and (b) average PSNR comparison

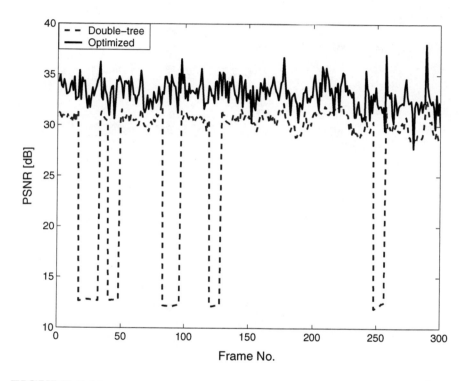

FIGURE 5.11
Frame PSNR comparison at node 10 between the optimized scheme and the double-tree scheme for video multicasting over a CDMA wireless ad hoc network

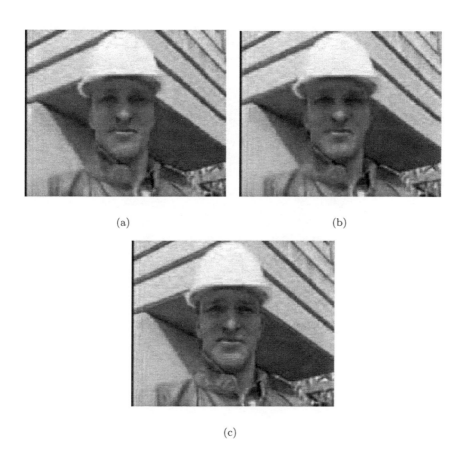

(a)

(b)

(c)

FIGURE 5.12
Comparison of reconstructed frame 13 of Foreman QCIF sequence at node
10 among three schemes for video multicasting over a CDMA wireless ad hoc
network: (a) the single-tree scheme (30.747 dB), (b) the double-tree scheme
(31.282 dB), and (c) the optimized scheme (34.755 dB)

5.5   Chapter Summary

In this chapter, we have studied the distributed optimization for video streaming over wireless ad hoc networks. We have employed a prioritized coding scheme which combines layered video coding and network coding. With a prioritized coding scheme, heterogeneous receivers can reconstruct the video at different levels of the video quality according to their capacities. The integration of the network coding eliminates the delivery redundancy, thus the receiver receives the distinct packets with very high probability. Video streaming can be classified into unicast streaming and multicast streaming. We have applied our distributed framework to both streaming classes. For unicast streaming over wireless ad hoc networks, we minimized the expected distortion by jointly optimizing both the source rate and the routing scheme. The numerical results showed that the proposed optimized routing can outperform the congestion-minimized routing and double-disjoint-path routing scheme. For the multicast streaming over wireless ad hoc networks, we examined the video streaming under different communication models including FDMA and CDMA. In video multicasting over FDMA wireless ad hoc networks, we optimized both the source rate and the routing scheme, and achieved a significant improvement in video quality over the double-tree routing scheme. In video multicasting over CDMA wireless ad hoc networks, we presented a distributed cross-layer optimization, taking into account the power allocation in the physical layer. Namely, we jointly optimized the source rate, the routing scheme and the power allocation for the video multicast using the hierarchical dual decompositions. Extensive simulations illustrated significant gains in terms of video quality over both the uniform-power scheme and the tree-based schemes.

Bibliography

[1] W. Wei and A. Zakhor, "Multipath unicast and multicast video communication over wireless ad hoc networks," in *Proc. of IEEE BroadNets,* pp. 496-505, Oct. 2004.

[2] Y. He, I. Lee, and L. Guan, "Optimized video multicasting over wireless ad hoc networks using distributed algorithm," *IEEE Transactions on Circuits and Systems for Video Technology,* vol. 19, no. 6, pp. 796-807, Jun. 2009.

[3] K. Stuhlmuller, N. Farber, M. Link, and B. Girod, "Analysis of video transmission over lossy channels," *IEEE Journal on Selected Areas in Communications,* vol. 18, no. 6, pp. 1012-1032, June 2000.

[4] R. Ahlswede, N. Cai, S.-Y. R. Li and R. W. Yeung, "Network information flow," *IEEE Transactions on Information Theory,* vol. 46, pp. 1204-1216, Jul. 2000.

[5] Z. Li, B. Li, D. Jiang, and L. C. Lau, "On Achieving Optimal Throughput with Network Coding," in *Proc. of IEEE INFOCOM,* vol. 3, pp. 2184-2194, Mar. 2005.

[6] S. Mao, S. Lin, S. Panwar, Y. Wang, and E. Celebi, "Video transport over ad hoc networks: Multistream coding with multipath transport," *IEEE Journal on Selected Areas in Communications,* vol. 21, no. 10, pp. 1721-1737, Dec. 2003.

[7] W. Wei and A. Zakhor, "Robust multipath source routing protocol (RMPSR) for video communication over wireless ad hoc networks," in *Proc. of IEEE ICME,* vol. 2, pp. 1379-1382, Jun. 2004.

[8] S. Mao, Y. T. Hou, X. Cheng, H. D. Sherali, and S. F. Midkiff, "Multipath routing for multiple description video over wireless ad hoc networks," in *Proc. of IEEE INFOCOM,* pp. 740-750, Mar. 2005.

[9] X. Zhu and B. Girod, "A distributed algorithm for congestion-minimized multi-path routing over ad hoc networks," in *Proc. of IEEE ICME,* pp. 1484-1487, Jul. 2005.

[10] D. Palomar and M. Chiang, "A tutorial on decomposition methods and distributed network resource allocation," *IEEE Journal on Selected Areas in Communications,* vol. 24, no. 8, pp. 1439-1451, Aug. 2006.

[11] S. Boyd and L. Vandenberghe, *Convex Optimization,* Cambridge University Press, 2004.

[12] D. P. Bertsekas, A. Nedic, and A. E. Ozdaglar, *Convex Analysis and Optimization,* Athena Scientific, 2003.

[13] Y. He, I. Lee, and L. Guan, "Optimized multi-path routing using dual decomposition for wireless video streaming," in *Proc. of IEEE ISCAS,* pp. 977-980, May 2007.

[14] H. Schwarz, D. Marpe, and T. Wiegand, "SNR-scalable extension of H.264/AVC," in *Proc. of IEEE ICIP,* vol. 5, pp. 3113-3116, Oct. 2004.

[15] Y. He, I. Lee, and L. Guan, "Video multicast over wireless ad hoc networks using distributed optimization," in *Proc. of Pacific-rim Conference on Multimedia (PCM),* pp. 296-305, Dec. 2007.

[16] S. Mao, S. Lin, S. Panwar, and Y. Wang, "Reliable transmission of video over ad-hoc networks using automatic repeat request and multi-path transport," in *Proc. of IEEE VTC,* vol. 2, pp. 615-619, Oct. 2001.

[17] P. Mohapatra, C. Gui, and J. Li, "Group communications in mobile ad hoc networks," *IEEE Computer,* vol. 37, no. 2, pp. 52-59, Feb. 2004.

[18] F. P. Kelly, A. Maulloo, and D. Tan, "Rate control for communication networks: shadow prices, proportional fairness and stability," *Journal of Operations Research Society,* vol. 49, no. 3, pp. 237-252, Mar. 1998.

[19] M. Chiang, S. H. Low, A. R. Calderbank, and J. C. Doyle, "Layering as optimization decomposition: A mathematical theory of network architectures," *Proceedings of the IEEE,* vol. 95, no. 1, pp. 255-312, Jan. 2007.

[20] J. Yuan, Z. Li, W. Yu, B. Li, "A cross-layer optimization framework for multihop multicast in wireless mesh networks," *IEEE Journal on Selected Areas in Communications,* vol. 24, no. 11, pp. 2096-2103, Nov. 2006.

[21] L. Xiao, M. Johansson and S. Boyd, "Simultaneous routing and resource allocation via dual decomposition," *IEEE Transactions on Communications,* vol. 52, no. 7, pp. 1136-1144, Jul. 2004.

[22] L. Chen, S. H. Low, M. Chiang, and J. C. Doyle, "Cross-layer congestion control, routing and scheduling design in ad hoc wireless networks," in *Proc. of IEEE INFOCOM,* pp. 1-13, Apr. 2006.

[23] X. Zhu, J. P. Singh, and B. Girod, "Joint routing and rate allocation for multiple video streams in ad hoc wireless networks," *Journal of Zhejiang University,* Science A, vol. 7, no. 5, pp. 727-736, May 2006.

[24] P. Papadimitratos, Z. Haas, and E. Sirer, "Path set selection in mobile ad hoc networks," in *Proc. of ACM MOBIHOC,* pp. 1-11, Jun. 2002.

[25] S. Mao, X. Cheng, Y. T. Hou, and H. D. Sherali, "Multiple description video multicast in wireless ad hoc networks," in *Proc. of IEEE BROAD-NETS,* pp. 671-680, Oct. 2004.

[26] A. Zakhor and W. Wei, "Multiple Tree Video Multicast over Wireless Ad Hoc Networks," in *Proc. of ICIP,* pp. 1665-1668, Sep. 2006.

6

Optimal Resource Allocation for Wireless Visual Sensor Networks

CONTENTS

Network lifetime maximization is a critical issue in Wireless Sensor Networks (WSNs) since each sensor has a limited power supply. In contrast with conventional sensor networks, video sensor nodes compress the video before transmission. The encoding process demands high power consumption, thus raising a challenge to the maintenance of a long network lifetime. In this chapter, we examine a strategy for maximizing the network lifetime in Wireless Visual Sensor Networks (WVSNs) by jointly optimizing the source rates, the encoding powers and the routing scheme.

6.1   Introduction

A wireless sensor network consists of geographically distributed sensors that communicate with each other over wireless channels [1]. A wireless visual sensor network is a special kind of WSN in that each sensor is equipped with video capture and processing components. WVSN facilitates a wide range of applications, such as video surveillance, emergency response, environmental tracking, and health monitoring [2].

Each video sensor in the WVSN has a camera component to capture the video, and a processing component to compress the video. The video sensors construct a mesh network topology, and they communicate with each other within a limited transmission range. The video captured and encoded at each sensor is transmitted to a sink for further analysis and decision making.

Sensor nodes are typically battery powered, and battery replacement is infrequent or even impossible in many sensing applications. Hence, a tremendous amount of research effort in wireless sensor networks has been focused on energy conservation. One aspect of this research is to maximize the network lifetime. In conventional wireless sensor networks, the data processing performed by the sensor node is assumed to be very simple. Thus, the energy consumption utilized for data collection and processing is often negligible. In contrast, video sensors in a WVSN need to compress the data prior to transmission. Efficient video compression algorithms are typically associated with high power consumption.

It is quite challenging to prolong or maximize the network lifetime for a WVSN. First, the algorithms, which maximize the network lifetime for conventional wireless sensor networks, focus on the allocation of transmission power and reception power. These algorithms cannot perform well when applied directly to WVSNs, since they neglect the power consumption on signal processing. Second, there is a tradeoff between the video quality and the network lifetime. A WVSN can extend its network lifetime by sacrificing the quality of the collected videos.

Network lifetime maximization for conventional wireless sensor networks has been extensively studied in the past. Chang et al. [3] developed a maximum lifetime routing scheme. Madan et al. [4] solved the lifetime maximization problem with a distributed algorithm using the dual decomposition and the subgradient method. Lifetime maximization for interference-limited networks using a cross-layer approach was studied in [5]. The tradeoff between the source rate allocation and the network lifetime was investigated in [6][7]. However, these methods [3][4][5][6][7] cannot be applied directly to the wireless visual sensor networks, since they omit the processing power consumption at the sensor nodes.

Wireless visual sensor networks have recently become an active research area. In [8], the concept of accumulative visual information was introduced as a means for measuring the amount of visual information collected in a WVSN. Minimizing the video distortion by optimizing the power allocation in WVSNs was investigated in [2]. He et al. investigated the resource-distortion optimization problem for video encoding and transmission over WVSNs [9]. A cooperative relaying architecture for delivering aggregated high-rate video data to the destination in wireless video sensor networks was proposed in [10]. In order to prolong the network lifetime for wireless sensor networks, an application-aware routing protocol, Distributed Activation based on Prede-termined Routes (DAPR) [11], was proposed by avoiding the use of sensors in the sparsely deployed areas as routers. Soro et al. extended DAPR in WVSNs [12] by introducing the total cost that combines the coverage and routing costs for each video sensor. However, DAPR cannot prolong the network lifetime for WVSNs to maximum, since the encoding power at each sensor node has not been optimized.

This chapter tackles the network lifetime maximization problem in wireless visual sensor networks. We present a distributed algorithm to maximize the network lifetime by jointly optimizing the source rates, the encoding powers, and the routing scheme. We investigate the network lifetime maximization problem in both large-delay applications and small-delay applications, respec-tively. In addition, we provide the relationship between the collected video quality and the maximal network lifetime in WVSNs.

The rest of the chapter is organized as follows. The system models for the WVSN are described in Section 6.2. In Section 6.3, we study the achiev-able maximum network lifetime in the WVSN without transmission errors. The network lifetime maximization in the WVSN with transmission errors is investigated in Section 6.4 and Section 6.5. Section 6.4 targets large-delay WVSN applications, while Section 6.5 targets small-delay WVSN applications. Finally, we summarize this chapter in Section 6.6.

6.2   System Models

In this section, we describe the network graph, the channel error model, the power consumption model, and give the definition of network lifetime. These models and definition will be used to formulate the network lifetime maxi-mization problem in the next sections.

6.2.1   Network Graph

A static wireless visual sensor network can be modeled as a directed graph $\mathbf{G} = (\mathbf{N}, \mathbf{L})$, where $\mathbf{N}$ is the set of video sensors and $\mathbf{L}$ is the set of directed wireless links. Among N nodes, one node belongs to the sink set $\mathbf{T}$, while the other nodes belong to video sensor set $\mathbf{V}$. Thus, $\mathbf{N} = \mathbf{V} \bigcup \mathbf{T}$. In this work, we assume that there is only one sink. However, the setup can be easily extended to include multiple sinks. Two nodes i and j are connected by a link if they can directly communicate with each other.

The relationship between a WVSN node and its connected links is represented with a node-link incidence matrix $\mathbf{A}$, whose elements are defined as

$$a_{il} = \begin{cases} 1, & \text{if link } l \text{ is an outgoing link from node } i, \\ -1, & \text{if link } l \text{ is an incoming link into node } i, \\ 0, & \text{otherwise.} \end{cases} \quad (6.1)$$

The relationship between a WVSN node and its outgoing links is represented with a matrix $\mathbf{A}^+$, whose elements are given by

$$a_{il}^+ = \begin{cases} 1, & \text{if link } l \text{ is an outgoing link from node } i, \\ 0, & \text{otherwise.} \end{cases} \quad (6.2)$$

The relationship between a WVSN node and its incoming links is represented with a matrix $\mathbf{A}^-$, whose elements are given by

$$a_{il}^- = \begin{cases} 1, & \text{if link } l \text{ is an incoming link from node } i, \\ 0, & \text{otherwise.} \end{cases} \quad (6.3)$$

Hence, $\mathbf{A} = \mathbf{A}^+ - \mathbf{A}^-$.

We assume a standard Medium Access Control (MAC) layer protocol is applied to resolve the link interference problem. Sensor node $h, \forall h \in \mathbf{V}$, can capture and encode the video, and then generate data traffic with a source rate s_h ($s_h = 0$ if sensor h is not on the capture and encoding mode). We define session h as the traffic flow originating from the sensor node h to the sink. For each session, the flow conservation law holds at each node:

$$\sum_{l \in \mathbf{L}} a_{il} x_{hl} = \eta_{hi}, \quad \forall h \in \mathbf{V}, \ \forall i \in \mathbf{N}, \quad (6.4)$$

where x_{hl} is the flow rate at link l for session h, and η_{hi} is defined as

$$\eta_{hi} = \begin{cases} s_h, & \text{if } i \text{ is the source node of sesion } h, \\ -s_h, & \text{if } i \text{ is the sink of sesion } h, \\ 0, & \text{otherwise.} \end{cases} \quad (6.5)$$

6.2.2 Channel Error Model

In wireless sensor networks, the channel at each link can be modeled as an Independent and Identically Distributed (IID) random bit error channel. Here, we employ a two-state Markov chain [13] to model the stochastic channel error pattern. The two states of the model are denoted as "1" (GOOD) and "0" (BAD). If the channel is in the GOOD state, the bit will be received correctly, and if the channel is in the BAD state, the bit will be received with error. The two-state Markov chain has been widely used to simulate the error patterns in wireless ad hoc networks [14] and wireless sensor networks [15].

Based on the Markov channel error model, the average bit error probability p_l^b at link l is then given by

$$p_l^b = \frac{q_l^{10}}{q_l^{10} + q_l^{01}}, \tag{6.6}$$

where q_l^{10} is the transition probability from a GOOD state to a BAD state, and q_l^{01} is the transition probability from a BAD state to a GOOD state.

Each packet has a fixed length of G bits. A packet is regarded as lost when any bit error in that packet occurs. Then the Packet Loss Rate (PLR) at link l is given by

$$p_l^p = 1 - (1 - p_l^b)^G. \tag{6.7}$$

6.2.3 Power Consumption Model

Video sensor h captures and encodes the video before it transmits the traffic to its downstream node. The distortion of the compressed video depends on the source rate s_h and the *encoding power consumption* P_{sh}. According to the Power–Rate–Distortion (P–R–D) analytical model in [16], the encoding distortion is computed by

$$d_{sh} = \sigma^2 e^{-\gamma \cdot s_h \cdot P_{sh}^{2/3}}, \tag{6.8}$$

where σ^2 is the input variance, γ is the encoding efficiency coefficient.

From the P–R–D model, we can see the following relationships: 1) At a fixed encoding power, the encoding distortion can be reduced by increasing the source rate; 2) At a fixed source rate, the encoding distortion can be reduced by increasing the encoding power. Thus, to achieve a given encoding distortion requirement for session h, we can either increase the encoding power or increase the source rate. However, increasing the encoding power may raise the power consumption of the source node. On the other hand, increasing the source rate may cause the downstream nodes to consume more power in order to relay extra traffic to the sink. Therefore, optimal allocation between the source rate and the encoding power is critical for maximizing network lifetime.

Based on a power consumption model in wireless sensor networks [17], the *transmission power consumption* at link l can be formulated as:

$$P_{tl} = c_l^s y_l, \text{ and } c_l^s = \alpha + \beta d_l^{n_p}, \tag{6.9}$$

where y_l is the aggregate rate transmitted through link l, c_l^s is the transmission energy consumption cost of link l, α is the energy cost of transmit electronics, β is a coefficient term corresponding to the energy cost of transmit amplifier, d_l is the distance between the transmitter and the receiver along link l, and n_p is the path-loss exponent [18].

The *reception power consumption* at a node i can be formulated as:

$$P_{ri} = c^r \sum_{l \in \mathbf{L}} a_{il}^- y_l, \tag{6.10}$$

where c^r is the energy consumption cost of the radio receiver, and $\sum_{l \in \mathbf{L}} a_{il}^- y_l$ represents the aggregate rate received at node i.

The total power dissipation at node i consists of *the encoding power consumption, the transmission power consumption* and *the reception power consumption,* and is given by

$$P_i = P_{si} + P_{ti} + P_{ri} = P_{si} + \sum_{l \in \mathbf{L}} a_{il}^+ (c_l^s y_l) + c^r \sum_{l \in \mathbf{L}} a_{il}^- y_l, \tag{6.11}$$

where $P_{si} = 0$, if i is not in the video sensor set $\mathbf{V}$.

6.2.4 Network Lifetime

Generally, the network lifetime is defined as the time period from the start time of the network until the time when the whole network fails due to the energy exhaustion of a set of sensors $\mathbf{K}$, where $\mathbf{K} \subseteq \mathbf{N}$. In some mission-critical applications, each sensor node is critical to the operation of the network operation. The exhaustion of energy of any node will cause the failure of the whole network. For example, each access is monitored by a visual sensor in a security-monitoring application. If any of the visual sensors fails due to energy exhaustion, the intruder can break into the monitored area. In that case, the whole security-monitoring system loses its function even though most of the visual sensors are working well. The network lifetime in such applications is defined as the minimum node lifetime [3][4][6]. In a WVSN, sensor node i has an initial energy B_i, and the lifetime of node i is given by $T_i = B_i/P_i, \forall i \in \mathbf{N}$. Then the network lifetime is given by $T_{net} = \min_{i \in \mathbf{N}} \{T_i\} = \min_{i \in \mathbf{N}} \{B_i/P_i\}$. In the following, we will use the minimum node lifetime as the network lifetime.

6.3 Achievable Maximum Network Lifetime

In wireless visual sensor networks, the sink collects each video from the corresponding sensor. The distortion of each video consists of an encoding distortion and a transmission distortion (distortion due to transmission errors). There is a tradeoff between the distortion requirement and the maximum network lifetime. If a high-quality video is desired, the maximum network lifetime will be compromised. On the other hand, WVSN can survive a longer time if it has a lower quality (larger distortion) requirement.

Given a distortion requirement at the sink, the maximum network lifetime depends on the network status. In a WVSN in the presence of transmission errors, sensor nodes can use various techniques, such as retransmissions, or Forward Error Correction (FEC), to combat the transmission errors. However, these methods introduce additional power consumption and thus reduce the network lifetime. On the other hand, if sensor nodes do not apply retransmission or FEC to recover from transmission errors, the transmission distortion will be introduced. Subject to the total distortion requirement, video sensors need to encode the video with a smaller encoding distortion, which correspondingly requires a higher encoding power or a larger source rate, thus reducing the network lifetime.

The achievable maximum network lifetime in a WVSN is obtained when there is no transmission error. In this section, we will investigate the achievable maximum network lifetime. The maximum network lifetime in a WVSN with transmission errors will be examined in the next two sections: Section 6.4 and Section 6.5.

6.3.1 Problem Formulation

In a WVSN that has no transmission error, the total distortion is equal to the encoding distortion since the transmission distortion is zero. In this case, the received video at the sink is measured by the encoding distortion in Mean Squared Error (MSE).

We state the problem of the achievable maximum network lifetime as: given the topology of a static WVSN, and the initial energy at each node, to maximize the network lifetime by jointly optimizing the source rate and the encoding power at each video sensor, and the link rate of each session, subject to an encoding distortion requirement for each captured video. Mathematically, the problem can be formulated as follows.

$$\text{maximize} \atop (\mathbf{s,x,P_s})} \quad \{T_{net} = \min\{B_i/(P_{si} + \textstyle\sum_{l\in\mathbf{L}} a_{il}^+(c_l^s y_l) + c^r \sum_{l\in\mathbf{L}} a_{il}^- y_l)\}\}$$

$$\text{subject to} \quad \begin{aligned}
&\textstyle\sum_{l\in\mathbf{L}} a_{il} x_{hl} = \eta_{hi}, \quad \forall h \in \mathbf{V}, \forall i \in \mathbf{N}, \\
&\textstyle\sum_{h\in\mathbf{V}} x_{hl} = y_l, \quad \forall l \in \mathbf{L}, \\
&\sigma^2 e^{-\gamma \cdot s_h \cdot P_{sh}^{2/3}} \le D_h, \quad \forall h \in \mathbf{V}, \\
&x_{hl} \ge 0, \quad \forall h \in \mathbf{V}, \forall l \in \mathbf{L}, \\
&s_h \ge 0, \quad \forall h \in \mathbf{V}, \\
&P_{sh} \ge 0, \quad \forall h \in \mathbf{V},
\end{aligned}$$

$$(6.12)$$

where B_i is the initial energy at node i, P_i is the total power consumption at node i, x_{hl} is the link rate at link l for session h, y_l is the aggregate flow rate at link l, s_h is the source rate of session h, P_{sh} is the encoding power at the source node of session h, and D_h is the upper bound of the encoding distortion for session h in MSE. The first constraint $\sum_{l\in\mathbf{L}} a_{il} x_{hl} = \eta_{hi}$ represents the flow conservation at each node for each session, the second constraint $\sum_{h\in\mathbf{V}} x_{hl} = y_l$ represents that the aggregate flow rate y_l at a link is the summation of the link rates of all the sessions at this link, and the third constraint $\sigma^2 e^{-\gamma \cdot s_h \cdot P_{sh}^{2/3}} \leq D_h$ represents that the encoding distortion for session h is required to be no larger than the corresponding upper bound D_h.

We replace the variable T_{net} using $q = 1/T_{net}$. Since $T_{net} \leq B_i/(P_{si} + \sum_{l \in \mathbf{L}} a_{il}^+(c_l^s y_l) + c^r \sum_{l \in \mathbf{L}} a_{il}^- y_l), \forall i \in \mathbf{N}$, we have $qB_i \geq P_{si} + \sum_{l \in \mathbf{L}} a_{il}^+(c_l^s y_l) + c^r \sum_{l \in \mathbf{L}} a_{il}^- y_l, \forall i \in \mathbf{N}$. Then the problem (6.12) is converted to an equivalent formulation as follows.

$$
\begin{aligned}
&\underset{(\mathbf{s},\mathbf{x},\mathbf{P_s},q)}{\text{minimize}} \quad q \\
&\text{subject to} \quad \sum_{l \in \mathbf{L}} a_{il} x_{hl} = \eta_{hi}, && \forall h \in \mathbf{V}, \forall i \in \mathbf{N}, \\
&\qquad\qquad \sum_{h \in \mathbf{V}} x_{hl} = y_l, && \forall l \in \mathbf{L}, \\
&\qquad\qquad \log(\sigma^2/D_h)/(\gamma P_{sh}^{2/3}) \leq s_h, && \forall h \in \mathbf{V}, \\
&\qquad\qquad P_{si} + \sum_{l \in \mathbf{L}} a_{il}^+(c_l^s y_l) + c^r \sum_{l \in \mathbf{L}} a_{il}^- y_l \leq qB_i, && \forall i \in \mathbf{N}, \\
&\qquad\qquad x_{hl} \geq 0, && \forall h \in \mathbf{V}, \forall l \in \mathbf{L}, \\
&\qquad\qquad s_h \geq 0, && \forall h \in \mathbf{V}, \\
&\qquad\qquad P_{sh} \geq 0, && \forall h \in \mathbf{V}.
\end{aligned}
\tag{6.13}
$$

The optimization problem (6.13) cannot be solved in a fully distributed manner, because the value of q needs to be broadcast to each node. In order to develop a fully distributed algorithm, we introduce an auxiliary variable $q_i, \forall i \in \mathbf{N}$ for node i. In problem (6.13), each node maintains a common q, which is equivalent to the case that node i maintains an individual q_i while $q_i = q_j, \forall i, j \in \mathbf{N}$. The equality constraint $q_i = q_j, \forall i, j \in \mathbf{N}$ can be represented in an another way $\sum_{i \in \mathbf{N}} a_{il} q_i = 0, \forall l \in \mathbf{L}$.

We know $q = (1/T_{net}) > 0$, the objective that minimizes q is equivalent to the one that minimizes $|\mathbf{N}|q^2$, where $|\mathbf{N}|$ represents the number of nodes in the WVSN. By using auxiliary variable q_i to replace the common q, the objective function $|\mathbf{N}|q^2$ is equal to $\sum_{i \in \mathbf{N}} q_i^2$ under the equality constraint $\sum_{i \in \mathbf{N}} a_{il} q_i = 0, \forall l \in \mathbf{L}$. Therefore, the optimization problem (6.13) is converted to the following formulation:

$$
\begin{aligned}
&\underset{(\mathbf{s},\mathbf{x},\mathbf{P_s},\mathbf{q})}{\text{minimize}} \quad \sum_{i \in \mathbf{N}} q_i^2 \\
&\text{subject to} \quad \sum_{l \in \mathbf{L}} a_{il} x_{hl} = \eta_{hi}, && \forall h \in \mathbf{V}, \forall i \in \mathbf{N}, \\
&\qquad\qquad \log(\sigma^2/D_h)/(\gamma P_{sh}^{2/3}) \leq s_h, && \forall h \in \mathbf{V}, \\
&\qquad\qquad P_{si} + \sum_{l \in \mathbf{L}} a_{il}^+(c_l^s \sum_{h \in \mathbf{V}} x_{hl}) + \\
&\qquad\qquad\quad c^r \sum_{l \in \mathbf{L}} a_{il}^- \sum_{h \in \mathbf{V}} x_{hl} \leq q_i B_i, && \forall i \in \mathbf{N}, \\
&\qquad\qquad \sum_{i \in \mathbf{N}} a_{il} q_i = 0, && \forall l \in \mathbf{L}, \\
&\qquad\qquad x_{hl} \geq 0, && \forall h \in \mathbf{V}, \forall l \in \mathbf{L}, \\
&\qquad\qquad q_i > 0, && \forall i \in \mathbf{N}, \\
&\qquad\qquad s_h \geq 0, && \forall h \in \mathbf{V}, \\
&\qquad\qquad P_{sh} \geq 0, && \forall h \in \mathbf{V},
\end{aligned}
\tag{6.14}
$$

In order to develop a fully distributed algorithm, we add a corresponding quadratic regularization term for each link rate variable and source rate variable to make the objective function strictly convex. The optimization problem (6.14) is approximated to:

$$
\begin{aligned}
\underset{(\mathbf{s},\mathbf{x},\mathbf{P_s},\mathbf{q})}{\text{minimize}} \quad & \sum_{i\in\mathbf{N}} q_i^2 + \sum_{h\in\mathbf{V}}\sum_{l\in\mathbf{L}} \delta x_{hl}^2 + \sum_{h\in\mathbf{V}} \delta s_h^2 \\
\text{subject to} \quad & \text{the same constraints as in (6.14)}
\end{aligned}
\tag{6.15}
$$

where $\delta(\delta > 0)$ is the regularization factor. When the regularization factor δ is close to 0, the objective value in problem (6.15) will be close to the objective value in problem (6.14). Therefore, the solution for the problem (6.15) is close to the solution for the original optimization problem (6.12).

6.3.2   Distributed Algorithm

In the problem (6.15), the objective function is strictly convex, the inequality functions are convex, and the equality functions are linear. Therefore, it is a convex optimization problem [19]. In addition, there exists a strictly feasible solution that satisfies all the constraints in the problem (6.15). In other words, the Slater's condition is satisfied, and the strong duality holds [19]. Thus, we can obtain the optimal solutions indirectly by first solving the corresponding dual problem [19]. The dual-based approach leads to an efficient distributed algorithm [20].

We introduce dual variables $u_{hi}, \forall h \in \mathbf{V}, \forall i \in \mathbf{N}$; $v_h, \forall h \in \mathbf{V}$; $\lambda_i, \forall i \in \mathbf{N}$; $w_l, \forall l \in \mathbf{L}$ to formulate the Lagrangian corresponding to primal problem (6.15) as below

$$
\begin{aligned}
& L(\mathbf{s},\mathbf{x},\mathbf{P_s},\mathbf{q},\mathbf{u},\mathbf{v},\lambda,\mathbf{w}) \\
={}& \sum_{i\in\mathbf{N}} q_i^2 + \sum_{h\in\mathbf{V}}\sum_{l\in\mathbf{L}} \delta x_{hl}^2 + \sum_{h\in\mathbf{V}} \delta s_h^2 \\
& + \sum_{h\in\mathbf{V}}\sum_{i\in\mathbf{N}} u_{hi}\left(\sum_{l\in\mathbf{L}} a_{il}x_{hl} - \eta_{hi}\right) \\
& + \sum_{h\in\mathbf{V}} v_h\left(\log(\sigma^2/D_h)/(\gamma P_{sh}^{2/3}) - s_h\right) \\
& + \sum_{i\in\mathbf{N}} \lambda_i\left(P_{si} + \sum_{l\in\mathbf{L}} a_{il}^+(c_i^s \sum_{h\in\mathbf{V}} x_{hl}) + c^r \sum_{l\in\mathbf{L}} a_{il}^- \sum_{h\in\mathbf{V}} x_{hl} - q_i B_i\right) \\
& + \sum_{l\in\mathbf{L}} w_l \sum_{i\in\mathbf{N}} a_{il} q_i \\
={}& \sum_{i\in\mathbf{N}}\left(q_i^2 + q_i\left(\sum_{l\in\mathbf{L}} a_{il}w_l - \lambda_i B_i\right)\right) \\
& + \sum_{h\in\mathbf{V}}\left(v_h \log(\sigma^2/D_h)/(\gamma P_{sh}^{2/3}) + \lambda_h P_{sh}\right) \\
& + \sum_{h\in\mathbf{V}}\sum_{l\in\mathbf{L}}\left(\delta x_{hl}^2 + x_{hl}(c_i^s \sum_{i\in\mathbf{N}} \lambda_i a_{il}^+ + c^r \sum_{i\in\mathbf{N}} \lambda_i a_{il}^- + \sum_{i\in\mathbf{N}} u_{hi}a_{il})\right) \\
& + \sum_{h\in\mathbf{V}}\left(\delta s_h^2 - v_h s_h - \sum_{i\in\mathbf{N}} u_{hi}\eta_{hi}\right).
\end{aligned}
\tag{6.16}
$$

The Lagrange dual function $G(\mathbf{u},\mathbf{v},\lambda,\mathbf{w})$ is the minimum value of the Lagrangian over primal variables $\mathbf{s},\mathbf{x},\mathbf{P_s},\mathbf{q}$, and it is given by

$$
G(\mathbf{u},\mathbf{v},\lambda,\mathbf{w}) = \min\{L(\mathbf{s},\mathbf{x},\mathbf{P_s},\mathbf{q},\mathbf{u},\mathbf{v},\lambda,\mathbf{w})\}.
\tag{6.17}
$$

The Lagrange dual problem corresponding to the primal problem (6.15) is then given by

$$
\begin{aligned}
\underset{(\mathbf{u},\mathbf{v},\lambda,\mathbf{w})}{\text{maximize}} \quad & G(\mathbf{u},\mathbf{v},\lambda,\mathbf{w}) \\
\text{subject to} \quad & \mathbf{v} \geq \mathbf{0}, \quad \lambda \geq \mathbf{0}.
\end{aligned}
\tag{6.18}
$$

At the kth iteration, we calculate the primal variables as follows:

1. q_i at node i:

$$q_i^{(k)} = \arg\min_{q_i > 0}(q_i^2 + q_i(\sum_{l \in \mathbf{L}} a_{il} w_l^{(k)} - \lambda_i^{(k)} B_i)), \forall i \in \mathbf{N}. \qquad (6.19)$$

2. The encoding power P_{sh} at video sensor h:

$$P_{sh}^{(k)} = \arg\min_{P_{sh} \geq 0}(v_h^{(k)} \log(\sigma^2/D_h)/(\gamma P_{sh}^{2/3}) + \lambda_h^{(k)} P_{sh}), \forall h \in \mathbf{V}. \quad (6.20)$$

3. The source rate s_h at video sensor h:

$$s_h^{(k)} = \arg\min_{s_h \geq 0}(\delta s_h^2 - v_h^{(k)} s_h - \sum_{i \in \mathbf{N}} u_{hi}^{(k)} \eta_{hi}), \forall h \in \mathbf{V}. \qquad (6.21)$$

4. The link rate x_{hl} at link l for session h:

$$x_{hl}^{(k)} = \arg\min_{x_{hl} \geq 0} \quad (\delta x_{hl}^2 + x_{hl}(c_l^s \sum_{i \in \mathbf{N}} \lambda_i^{(k)} a_{il}^+ + c^r \sum_{i \in \mathbf{N}} \lambda_i^{(k)} a_{il}^- + \sum_{i \in \mathbf{N}} u_{hi}^{(k)} a_{il})), \quad \forall h \in \mathbf{V}, \forall l \in \mathbf{L}.$$

$$(6.22)$$

We use subgradient method [21] to solve the Lagrange dual problem. The dual variable at the $(k+1)^{th}$ iteration is updated by

$$u_{hi}^{(k+1)} = u_{hi}^{(k)} - \theta^{(k)}(\eta_{hi}^{(k)} - \sum_{l \in \mathbf{L}} a_{il} x_{hl}^{(k)}), \quad \forall h \in \mathbf{V}, \forall i \in \mathbf{N}, \qquad (6.23)$$

$$v_h^{(k+1)} = \max\{0, v_h^{(k)} - \theta^{(k)}(s_h^{(k)} - \log(\sigma^2/D_h)/(\gamma(P_{sh}^{(k)})^{2/3}))\}, \quad \forall h \in \mathbf{V}, \qquad (6.24)$$

$$\lambda_i^{(k+1)} = \max \quad \{0, \lambda_i^{(k)} - \theta^{(k)}(q_i^{(k)} B_i - \sum_{l \in \mathbf{L}} a_{il}^+ c_l^s \sum_{h \in \mathbf{V}} x_{hl}^{(k)} - c^r \sum_{l \in \mathbf{L}} a_{il}^- \sum_{h \in \mathbf{V}} x_{hl}^{(k)} - P_{si}^{(k)})\}, \forall i \in \mathbf{N}, \qquad (6.25)$$

$$w_l^{(k+1)} = w_l^{(k)} + \theta^{(k)} \sum_{i \in \mathbf{N}} a_{il} q_i^{(k)}, \quad \forall l \in \mathbf{L}, \qquad (6.26)$$

where $\theta^{(k)} > 0$ is the step size at k^{th} iteration. The step size we use in our algorithm is: $\theta^{(k)} = \omega/\sqrt{k}$, where $\omega > 0$.

The above algorithm is fully distributed. Each node computes the primal variables: 1) the auxiliary variable q_i, 2) the encoding power P_{sh}, 3) the source rate s_h, and 4) the outgoing link rate x_{hl} from this node, using the dual variables of itself and its neighboring nodes. When the dual variables converge, the primal variables also converge to their optimal values.

Parameter	Description	Value
σ^2	Input video variance	3500
γ	Encoding efficiency coefficient	$55.54W^{3/2}/Mbps$
α	Energy cost of transmit electronics	$0.5J/Mb$
β	Coefficient term of the transmit amplifier	$1.3 \times 10^{-8}J/Mb/m^4$
n_p	Path-loss exponent	4
c^r	Energy consumption cost of radio receiver	$0.5J/Mb$
δ	Regularization factor	0.2
ω	Step size parameter	0.15

TABLE 6.1
Configuration of model parameters in a wireless visual sensor network

6.3.3 Simulation Results

In this subsection, we evaluate the proposed distributed solution for the network lifetime maximization problem in the lossless scenario. We consider a static WVSN with 10 nodes randomly located in a square region of 50m-by-50m. Node 10 is the sink, and the other nodes are video sensors. Each node has a maximum transmission range of 30m. Foreman Common Intermediate Format (CIF) sequence is used in the simulations. The values of the model parameters are listed in Table 6.1. Each node has an initial energy of 5.0 MJ. All 9 video sensors encode the videos and transmit them to the sink. If not specified particularly, the upper bound of the encoding distortion D_h is set to 100, corresponding to a Peak Signal-to-Noise Ratio (PSNR) of 28.13 dB.

We introduce a regularization factor in order to derive the distributed algorithm. There is a tradeoff between the suboptimality and the computation complexity for different regularization factors, as shown in Figure 6.1. When the regularization factor is 0.01, we can obtain a maximum network lifetime of 1.02×10^7 s at the price of 7897 iterations. On the other hand, we can reduce the complexity to 277 iterations by sacrificing the maximum network lifetime to 8.69×10^6 s. In our simulations, we choose a regularization factor of 0.2, leading to a network lifetime of 9.41×10^6 s. Compared to the optimization with the regularization factor 0.01, the algorithm with the regularization factor 0.2 requires only 1/20 of the complexity though the maximum network lifetime is reduced only by 7.5%.

The proposed algorithm optimizes both the source (e.g., source rate and encoding power) and the routing scheme. We compare the proposed algorithm with two other schemes: 1) the Source-Optimized Scheme (SOS), in which only the encoding power and the source rate are optimized, while the routing is pre-determined with the shortest-path scheme; 2) the Routing-Optimized Scheme (ROS), in which only the routing scheme is optimized, while the en-

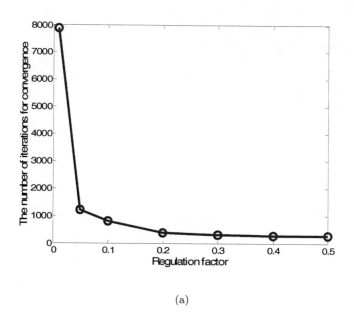

(a)

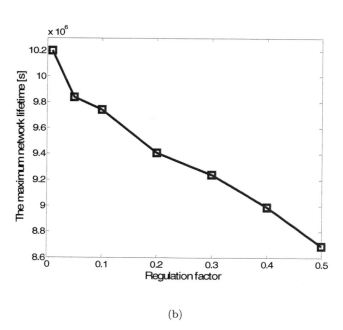

(b)

FIGURE 6.1
Tradeoff between the suboptimality and the complexity for different regularization factors: (a) the number of iterations for convergence, and (b) the maximum network lifetime

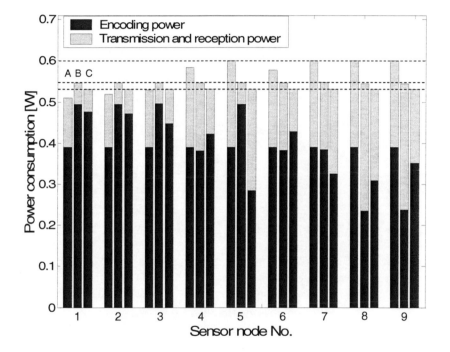

FIGURE 6.2
Comparison of power consumption at each sensor node

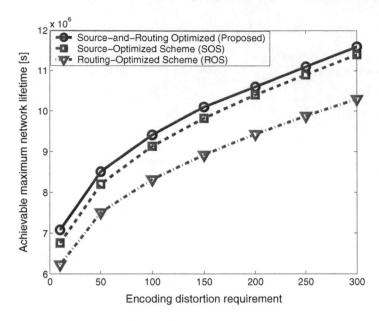

FIGURE 6.3
Tradeoff between encoding distortion requirement and achievable maximum
network lifetime in lossless transmission

coding power at each sensor node is fixed at the same value. In order for fair
comparison, the total encoding power of all the sensor nodes in the ROS is
equal to that in the proposed optimized scheme. The comparison of the power
consumption at each sensor node is illustrated in Figure 6.2. In the proposed
scheme, the power consumption at each sensor node (represented in bar C)
converges to the same level, meaning that all the sensor nodes will die at almost the same time. The power consumption in the ROS (represented in bar
A) and in the SOS (represented in bar B) at different sensor nodes is uneven,
thus some nodes will die before other nodes. The network lifetime is determined by the maximum power consumption among all the sensor nodes. The
maximum power consumption in the proposed algorithm is 0.5313 W, smaller
than that in ROS by 13.0% and SOS by 3.1%.

There is a tradeoff between the encoding distortion requirement and the
achievable maximum network lifetime. If a visual sensor network desires a
high-quality video, it will have to sacrifice its network lifetime. On the other
hand, a sensor network expecting a longer lifetime has to lower the quality
of the collected video. As shown in Figure 6.3, the proposed algorithm supports a longer network lifetime for different encoding distortion requirements,
compared to both the SOS and ROS.

6.4 Maximum Network Lifetime for Large-Delay Applications

The WVSN applications can be classified into two categories according to their delay requirement [2]. The first category is large-delay WVSN application, in which there is no stringent delay requirement. It only requires that the data be successfully delivered to the sink for future analysis. Environmental data collection belongs to this category. The second category is small-delay WVSN application, in which the video data is required to be transmitted to the sink, over the sensor networks, with a small delay for fast response and decision making. A real-time traffic monitoring system belongs to the second category.

In Section 6.3, we studied the achievable maximum network lifetime without transmission errors. If transmission errors exist, there is a reduction in the maximum network lifetime compared to the achievable maximum network lifetime. In this section, we investigate this reduction for the large-delay WVSN applications. We will examine the network lifetime reduction for small-delay WVSN applications in the next section.

6.4.1 Problem Formulation and Solution

Large-delay WVSN applications, such as video surveillance, typically maintain the video quality at a high priority. Corrupted packets can be retransmitted since delay is permitted. Retransmissions improve the overall video quality at the receiver. However, retransmissions also consume more energy, which will reduce the maximum network lifetime. Thus, we analyze the tradeoff between the power consumption and the reliability, and the impact on the maximum network lifetime.

At link l, the transmitter sends a packet to the receiver. If the packet is received correctly, the receiver will not notify the transmitter. On the other hand, if the packet is received with errors, the receiver will send back a Negative Acknowledge (NACK) to the transmitter to request retransmission. In this work, the NACK is kept simple and its energy consumption is assumed negligible. We also assume that NACK is always received correctly, and the maximum number of the retransmission N_{max} is sufficiently large. Using the Markov channel error model, the average number of transmissions $\overline{N_l}$ required over link l for successfully transmitting a packet is given by

$$\overline{N_l} = \sum_{k=0}^{N_{max}} (k+1)(p_l^p)^k(1 - p_l^p) \approx \frac{1}{1 - p_l^p}, \quad \forall l \in \mathbf{L}, \tag{6.27}$$

where p_l^p is the packet loss rate at link l, as defined in Equation (6.7).

In the retransmission scenario, the power consumption at node i is modified from Equation (6.11) to the following:

$$P_i = P_{si} + P_{ti} + P_{ri} = P_{si} + \sum_{l \in \mathbf{L}} a_{il}^+ \overline{N}_l (c_l^s \sum_{h \in \mathbf{V}} x_{hl}) + c^r \sum_{l \in \mathbf{L}} \overline{N}_l (a_{il}^- \sum_{h \in \mathbf{V}} x_{hl}), \ \forall i \in \mathbf{N}.$$
(6.28)

The network lifetime maximization problem with retransmissions for large-delay WVSN applications is modified from optimization problem (6.15) into the following form:

$$
\begin{aligned}
& \underset{(\mathbf{s},\mathbf{x},\mathbf{P_s},\mathbf{q})}{\text{minimize}} && \sum_{i \in \mathbf{N}} q_i^2 + \sum_{h \in \mathbf{V}} \sum_{l \in \mathbf{L}} \delta x_{hl}^2 + \sum_{h \in \mathbf{V}} \delta s_h^2 && \\
& \text{subject to} && \sum_{l \in \mathbf{L}} a_{il} x_{hl} = \eta_{hi}, && \forall h \in \mathbf{V}, \forall i \in \mathbf{N}, \\
& && \log(\sigma^2/D_h)/(\gamma P_{sh}^{2/3}) \le s_h, && \forall h \in \mathbf{V}, \\
& && P_{si} + \sum_{l \in \mathbf{L}} a_{il}^+ \overline{N}_l (c_l^s \sum_{h \in \mathbf{V}} x_{hl}) + && \\
& && \quad c^r \sum_{l \in \mathbf{L}} \overline{N}_l (a_{il}^- \sum_{h \in \mathbf{V}} x_{hl}) \le q_i B_i, && \forall i \in \mathbf{N}, \\
& && \sum_{i \in \mathbf{N}} a_{il} q_i = 0, && \forall l \in \mathbf{L}, \\
& && x_{hl} \ge 0, && \forall h \in \mathbf{V}, \forall l \in \mathbf{L}, \\
& && q_i > 0, && \forall i \in \mathbf{N}, \\
& && s_h \ge 0, && \forall h \in \mathbf{V}, \\
& && P_{sh} \ge 0, && \forall h \in \mathbf{V}.
\end{aligned}
$$
(6.29)

The network lifetime maximization problem in (6.29) is essentially the same as the achievable maximum network lifetime problem, except that the power consumption constraint in (6.29) integrates the retransmission power consumption. We can use the primal-dual method proposed in Section 6.3 to obtain a fully distributed solution. Simulation results for the retransmission case for large-delay WVSN applications are presented below.

6.4.2   Simulation Results

We use the same setup for the video sensor network and the same model parameters as in Section 6.3.3. In the simulation, we set the packet size to 512 bits. By varying the transition probability from a GOOD state to a BAD state q_l^{10} and the transition probability from a BAD state to a GOOD state q_l^{01}, we obtain different average PLRs.

In the error scenarios, each node needs to retransmit the corrupted packets, thus introducing more power consumption in transmission and reception. Extra power consumption leads to a reduction of network lifetime, which is shown in Figure 6.4. Compared to the lossless transmission, the maximum network lifetime is reduced by averagely 4.9% when the average PLR is 7.2%, 8.0% when the average PLR is 13.4% and 15.0% when the average PLR is 26.4%, respectively.

We compare the proposed scheme with the other two schemes: SOS and ROS. In Figure 6.5(a), we compare the maximum network lifetime at different encoding distortion requirements while the average PLR is 13.4%. In Figure 6.5(b), we set the encoding distortion requirement to 100, and compare the

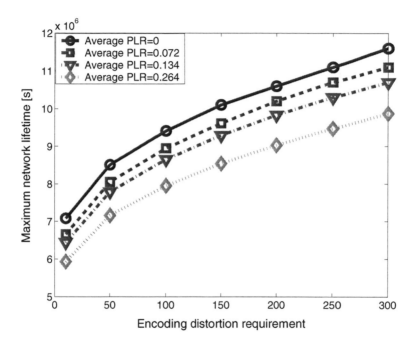

FIGURE 6.4

Reduction of maximum network lifetime in large-delay WVSN applications

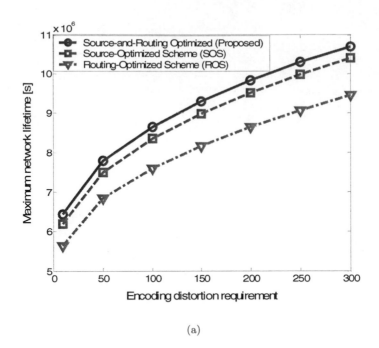

(a)

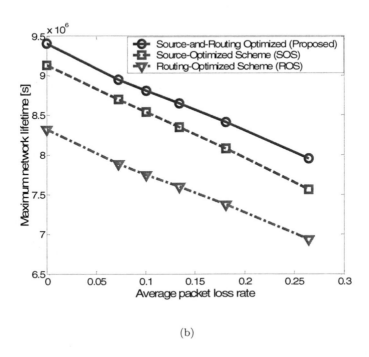

(b)

FIGURE 6.5
Comparison of maximum network lifetime in large-delay WVSN applications:
(a) with different encoding distortion requirement, and (b) with different average PLR.

maximum network lifetime by varying the average PLR. In both cases, the proposed algorithm optimizes the source coding and the routing scheme simultaneously, thus achieving a longer network lifetime compared to the SOS and the ROS. We also observe that SOS performs better than ROS. This is because that source encoding power takes a major part of the total power consumption, and therefore optimization of the source power utilization leads to a better performance over ROS.

6.5 Maximum Network Lifetime for Small-Delay Applications

For large-delay WVSN applications, wireless video sensors are allowed to retransmit the packets if they are not received correctly. However, for small-delay WVSN applications, due to the stringent delay requirement, packet retransmissions are infeasible even if the packets are received with errors.

In this section, we study the maximum network lifetime with and without FEC for small-delay WVSN applications.

6.5.1 With FEC

In wireless sensor networks, there are two popular methods to address the issues of corrupted or lost packets: Automatic Repeat Requests (ARQ) and FEC. ARQ requires packet retransmissions, which are typically applied for large-delay applications. FEC embeds additional bits to detect and recover from corrupted or lost data. This is suitable for small-delay sensing applications [15, 22].

The redundant bits introduced by FEC consume transmission power and reception power. Moreover, the encoding and decoding process of FEC also consumes additional power. Therefore, FEC reduces the maximum network lifetime compared to the achievable maximum network lifetime.

6.5.1.1 Problem Formulation and Solution

In this work, we use a kind of FEC, Reed-Solomon (RS) code [23] to recover the corrupted information in erroneous packets. RS code is widely used for image or video communications. Let $RS(n, k)$ be the code for transmission, where n is the block size in number of packets, $k(k < n)$ is the number of information packets, $m = n - k$ is the number of the redundancy packets. Any combination of $t_c = \lfloor (n - k)/2 \rfloor$ erroneous packets out of n can be recovered [15].

We use RS code at each link. Each node receives the RS blocks from its upstream nodes. The receiving node decodes the RS blocks and then re-encodes them before transmitting the RS codes to its downstream nodes. We apply Real-time Transport Protocol/Real-time Transport Control Protocol (RTP/RTCP) [24] at each node. Thus, each node can estimate the PLRs at its outgoing links. The packets are RS encoded adaptively according to the estimated PLR values.

We use two-state Markov chain to model the transmission channel, and we assume that the channel error at each link is independent from the amount of the traffic. RS coding is applied at each link. At link l, the packet loss rate is p_l^p, the information bit rate is x_{hl}, and the RS code bit rate is denoted by z_{hl}. The packet length is G bits/packet. Thus, the information rate in packets is $\lceil x_{hl}/G \rceil$ packets/second, and the rate of the RS code is $\lceil z_{hl}/G \rceil$ packets/second. In order to correct the erroneous packets, the number of erroneous packets needs to be less than or equal to the correction capacity of the RS code, which is expressed by

$$p_l^p \lceil z_{hl}/G \rceil \leq \lfloor \frac{\lceil z_{hl}/G \rceil - \lceil x_{hl}/G \rceil}{2} \rfloor. \tag{6.30}$$

We define a slack factor $\kappa (\kappa > 1)$. The bit rate of the RS code at link l is then given by $z_{hl} = x_{hl}/(1 - 2\kappa p_l^p)$. The slack factor is a given system parameter. A larger κ means a stronger protection for the information bits.

We use the power consumption model of RS codec proposed in [25]. The power consumption cost in RS encoding is c^{RSE}, and the power consumption cost in RS decoding is c^{RSD}. The power consumed by RS encoding at node i is given by

$$p_i^{RSE} = c^{RSE} \sum_{l \in \mathbf{L}} (a_{il}^+ \sum_{h \in \mathbf{V}} x_{hl}), \ \forall i \in \mathbf{N}. \tag{6.31}$$

The power consumed by RS decoding at node i is given by

$$p_i^{RSD} = c^{RSD} \sum_{l \in \mathbf{L}} (a_{il}^- \sum_{h \in \mathbf{V}} x_{hl}), \ \forall i \in \mathbf{N}. \tag{6.32}$$

The total power consumption at node i consists of *video encoding power consumption, RS encoding power consumption, transmission power consumption, RS decoding power consumption,* and *reception power consumption,* such that

$$\begin{aligned} P_i &= P_{si} + P_i^{RSE} + P_{ti} + P_i^{RSD} + P_{ri} \\ &= P_{si} + c^{RSE} \sum_{l \in \mathbf{L}} (a_{il}^+ \sum_{h \in \mathbf{V}} x_{hl}) + \sum_{l \in \mathbf{L}} a_{il}^+ (c_l^s \sum_{h \in \mathbf{V}} z_{hl}) + \\ &\quad c^{RSD} \sum_{l \in \mathbf{L}} (a_{il}^- \sum_{h \in \mathbf{V}} x_{hl}) + c^r \sum_{l \in \mathbf{L}} (a_{il}^- \sum_{h \in \mathbf{V}} z_{hl}) \\ &= P_{si} + c^{RSE} \sum_{l \in \mathbf{L}} (a_{il}^+ \sum_{h \in \mathbf{V}} x_{hl}) + \sum_{l \in \mathbf{L}} a_{il}^+ (c_l^s \sum_{h \in \mathbf{V}} \frac{x_{hl}}{1 - 2\kappa p_l^p}) + \\ &\quad c^{RSD} \sum_{l \in \mathbf{L}} (a_{il}^- \sum_{h \in \mathbf{V}} x_{hl}) + c^r \sum_{l \in \mathbf{L}} (a_{il}^- \sum_{h \in \mathbf{V}} \frac{x_{hl}}{1 - 2\kappa p_l^p}). \end{aligned} \tag{6.33}$$

The network lifetime maximization problem with RS code for small-delay WVSN applications is then modified from the optimization problem (6.15) into:

$$\underset{(\mathbf{s},\mathbf{x},\mathbf{P_s},\mathbf{q})}{\text{minimize}} \quad \sum_{i \in \mathbf{N}} q_i^2 + \sum_{h \in \mathbf{V}} \sum_{l \in \mathbf{L}} \delta x_{hl}^2 + \sum_{h \in \mathbf{V}} \delta s_h^2$$

$$
\begin{aligned}
\text{subject to} \quad & \sum_{l \in \mathbf{L}} a_{il} x_{hl} = \eta_{hi}, & & \forall h \in \mathbf{V}, \forall i \in \mathbf{N}, \\
& \log(\sigma^2/D_h)/(\gamma P_{sh}^{2/3}) \leq s_h, & & \forall h \in \mathbf{V}, \\
& P_{si} + c^{RSE} \sum_{l \in \mathbf{L}} (a_{il}^+ \sum_{h \in \mathbf{V}} x_{hl}) + \\
& \quad \sum_{l \in \mathbf{L}} a_{il}^+ (c_l^s \sum_{h \in \mathbf{V}} \tfrac{x_{hl}}{1-2\kappa p_l^p}) + \\
& \quad c^{RSD} \sum_{l \in \mathbf{L}} (a_{il}^- \sum_{h \in \mathbf{V}} x_{hl}) + \\
& \quad c^r \sum_{l \in \mathbf{L}} (a_{il}^- \sum_{h \in \mathbf{V}} \tfrac{x_{hl}}{1-2\kappa p_l^p}) \leq q_i B_i, & & \forall i \in \mathbf{N}, \\
& \sum_{i \in \mathbf{N}} a_{il} q_i = 0, & & \forall l \in \mathbf{L}, \\
& x_{hl} \geq 0, & & \forall h \in \mathbf{V}, \forall l \in \mathbf{L}, \\
& q_i > 0, & & \forall i \in \mathbf{N}, \\
& s_h \geq 0, & & \forall h \in \mathbf{V}, \\
& P_{sh} \geq 0, & & \forall h \in \mathbf{V}.
\end{aligned}
$$

$$(6.34)$$

The network lifetime maximization problem with RS code in (6.34) represents a convex problem which can be solved using a fully distributed algorithm via dual decomposition. The simulation results are presented below.

6.5.1.2   Simulation Results

We use the same model parameters shown in Table 6.1 and the error scenarios described in Section 6.4.2. The RS encoding power consumption cost is $c^{RSE} = 8 \times 10^{-5} J/Mb$, and the RS decoding power consumption cost is $c^{RSD} = 2.1 \times 10^{-4} J/Mb$. The slack factor κ is set to 1.1.

Figure 6.6 shows the reduction of network lifetime due to the extra power consumption at the RS coding and decoding. At the encoding distortion requirement of 100, the maximum network lifetime in the lossless transmission is 9.41×10^6 s. It is reduced by 9.6% at the average PLR of 7.2%, and 33.0% at the average PLR of 26.4%.

In the lossy scenarios with FEC, the proposed optimized scheme performs better than both the SOS and ROS. The comparison is demonstrated in Figure 6.7. The proposed scheme improves the network lifetime by 6.5%-7.7% over the SOS, and 14.8% -15.1% over the ROS, when the encoding distortion requirement is varied from 10 to 300. In Figure 6.7(b), we vary the PLR from 0 to 26.4%, the proposed scheme prolongs the network lifetime by 3.1%-25.1% over the SOS, and 13.1%-16.1% over the ROS. The improvement comes from the optimal power allocation in the source encoding, the RS encoding and decoding, the transmission and the reception.

6.5.2   Without FEC

For small-delay WVSN applications, if no FEC is used, transmission errors will cause decoding errors for the reconstructed video. This error may propagate over subsequent video frames due to drifting error, which exists in popular

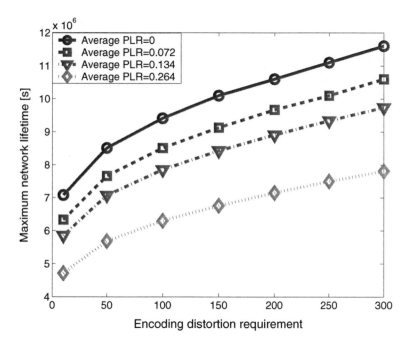

FIGURE 6.6

Reduction of maximum network lifetime in small-delay WVSN applications with FEC

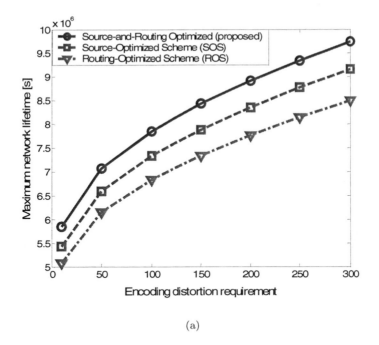

(a)

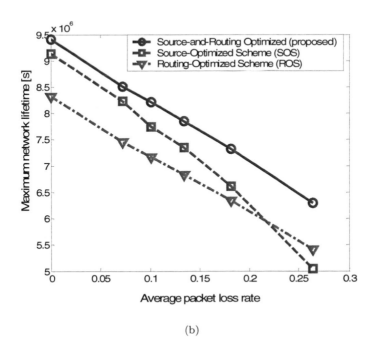

(b)

FIGURE 6.7

Comparison of maximum network lifetime in small-delay WVSN applications with FEC: (a) with different encoding distortion requirement, and (b) with different average PLR.

video coding techniques using block-based motion compensation. The video distortion caused by the transmission errors is called transmission distortion.

6.5.2.1 Transmission Distortion

The transmission distortion of each session depends on the end-to-end PLR. First we formulate the end-to-end PLR for each session, based on the Markov channel error model described in Section 6.2.2.

For session h, suppose there are J_h paths, labeled with $j_h = 1, ..., J_h$, from the video sensor h to the sink. Let $F(j_h)$ denote the set of the links in path j_h. Path j_h transports a normalized portion b_{j_h} of the source traffic s_h originating from the video sensor h, Hence, $b_1 + b_2 +, ..., + b_{J_h} = 1$. We define a binary variable for each link l:

$$Q_l^{j_h} = \begin{cases} 1, & \text{if } l \in F(j_h), \\ 0, & \text{otherwise.} \end{cases} \tag{6.35}$$

For path j_h, the end-to-end PLR can be computed as

$$p_{j_h}^E = 1 - \prod_{l \in \mathbf{L}}(1 - Q_l^{j_h} p_l^p). \tag{6.36}$$

Typically, the PLR p_l^p at each link is very small (e.g. $p_l^p \ll 1$). Thus, the end-to-end packet loss rate at path j_h can be approximated as $p_{j_h}^E \approx \sum_{l \in \mathbf{L}}(Q_l^{j_h} p_l^p)$ [26].

With multi-path routing, the traffic of session h is disseminated over J_h paths. Therefore, the end-to-end PLR for session h is given by

$$p_h^E = \sum_{j_h} b_{j_h} p_{j_h}^E \approx \sum_{j_h} b_{j_h} \sum_{l \in \mathbf{L}}(Q_l^{j_h} p_l^p) = \sum_{l \in \mathbf{L}} p_l^p \sum_{j_h}(b_{j_h} Q_l^{j_h}) = \frac{1}{s_h} \sum_{l \in \mathbf{L}} p_l^p x_{hl}. \tag{6.37}$$

We use the transmission distortion model presented in [27]. The transmission distortion d_{th} of session h is given by

$$d_{th} = \psi_h \frac{p_h^E}{1 - p_h^E} \overline{F_h} = \psi_h \frac{\frac{1}{s_h}\sum_{l \in \mathbf{L}} p_l^p x_{hl}}{1 - \frac{1}{s_h}\sum_{l \in \mathbf{L}} p_l^p x_{hl}} \overline{F_h} = \psi_h \overline{F_h} \frac{\sum_{l \in \mathbf{L}} p_l^p x_{hl}}{s_h - \sum_{l \in \mathbf{L}} p_l^p x_{hl}}, \tag{6.38}$$

where ψ_h is the model parameter, and $\overline{F_h}$ is the time average of the frame difference [27].

The total video distortion d_h of session h consists of the encoding distortion d_{sh} and the transmission distortion d_{th}. The former study concluded that the encoding distortion and transmission distortion are uncorrelated [27]. Therefore $d_h = d_{sh} + d_{th}$, where d_{sh} is given by Equation (6.8), and d_{th} is given by Equation (6.38).

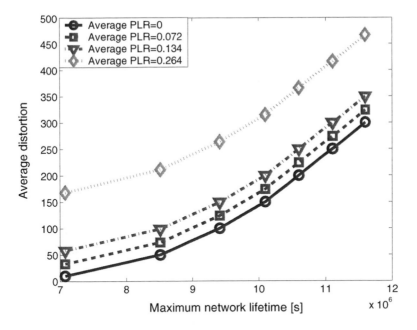

FIGURE 6.8
Tradeoff between the maximum network lifetime and the distortion in the lossy transmission

6.5.2.2 Tradeoff between Maximum Network Lifetime and Video Quality

We apply the distributed algorithm presented in Section 6.3 to the WVSN with transmission errors. The total distortion consists of the encoding distortion and the transmission distortion. Therefore, for the same network lifetime, the total video distortion with transmission errors is larger than that without transmission errors.

We present a simulation to evaluate the tradeoffs between the maximum network lifetime and the video distortion. The same model parameters in Table 6.1 are used in this simulation. The parameters of transmission distortion model are set as: $\psi_h = 0.8$ and $\overline{F_h} = 200$. As illustrated in Figure 6.8, the maximum network lifetime is increased at the price of the video distortion. At the same maximum lifetime, the packet errors introduce transmission distortion causing a larger average distortion compared to the lossless case. At the network lifetime of 9.41×10^6 s, we can receive a video at the distortion of 100 if there is no transmission error. The packet loss introduces transmission distortion, which degrades the received video quality. If the PLR is 7.2%, the distortion of the received video will be 123.8 under the same network lifetime 9.41×10^6 s. The distortion is increased to 263.4 if the PLR is increased to 26.4%.

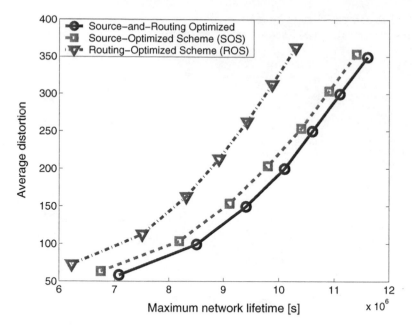

FIGURE 6.9

Comparison of three schemes in the lossy transmission

In Figure 6.9, we compare the tradeoff between the network lifetime and the distortion among the proposed scheme, the SOS, and the ROS. We can observe that the tradeoff trace in the proposed scheme is lower than those in the SOS and ROS. For the same network lifetime, the proposed scheme enables the sink to collect the video at a higher quality. In other words, the proposed scheme extends the network lifetime subject to the same distortion requirement.

6.6   Chapter Summary

In this chapter, we have studied the network lifetime maximization problem in wireless visual sensor networks. First, we investigated the achievable maximum network lifetime in WVSN without transmission errors. Then, we further examined the maximum network lifetime considering transmission errors. We investigated the error remedy techniques in both large-delay WVSN applications and small-delay WVSN applications, and studied their impacts on maximum network lifetime. We have derived distributed algorithms by using dual decomposition to maximize the network lifetime by jointly optimizing the source rates, the encoding powers, and the routing scheme.

Bibliography

[1] I. Akyildiz, W. Su, Y. Sankarasubramaniam, and E. Cayirci, "A survey on sensor networks," *IEEE Communication Magazine,* no. 8, pp. 102-114, Aug. 2002.

[2] Z. He and D. Wu, "Resource allocation and performance analysis of wireless video sensors," *IEEE Transactions on Circuits and Systems for Video Technology,* vol. 16, no. 5, pp. 590-599, May 2006.

[3] J. H. Chang and L. Tassiulas, "Maximum lifetime routing in wireless sensor networks," *IEEE/ACM Transactions on Networking,* vol. 12, no. 4, pp. 609-619, Aug. 2004.

[4] R. Madan, S. Lall, "Distributed algorithms for maximum lifetime routing in wireless sensor networks," *IEEE Transactions on Wireless Communications,* vol. 5, no. 8, pp. 2185-2193, Aug. 2006.

[5] R. Madan, S. Cui, S. Lall, and A. Goldsmith, "Cross-layer design for lifetime maximization in interference-limited wireless sensor networks," in *Proc. of IEEE INFOCOM,* vol. 3, pp. 1964-1975, Mar. 2005.

[6] J. Zhu, K. Hung, B. Bensaou, and F. Abdesselam, "Tradeoff between network lifetime and fair rate allocation in wireless sensor networks with multi-path routing," in *Proc. of ACM MSWiM,* pp. 301-308, Oct. 2006.

[7] H. Nama, M. Chiang, and N. Mandayam, "Utility-lifetime trade-off in self-regulating wireless sensor networks: A cross-layer design approach," in *Proc. of IEEE ICC,* vol. 8, pp. 3511-3516, Jun. 2006.

[8] Z. He and D. Wu, "Accumulative visual information in wireless video sensor network: definition and analysis," in *Proc. of IEEE ICC,* vol. 2, pp. 1205-1208, May 2005.

[9] Z. He and C. W. Chen, "From rate-distortion analysis to resource-distortion analysis," in *Proc. of IEEE WIRELESSCOM,* vol. 2, pp. 1527-1532, Jun. 2005.

[10] S. W. Kim, "Cooperative relaying architecture for wireless video sensor networks," in *Proc. of IEEE GLOBECOM,* vol. 5, pp. 3053-3057, Nov. 2005.

[11] M. Perillo and W. Heinzelman, "DAPR: a protocol for wireless sensor networks utilizing an application-based routing cost," in *Proc. of IEEE WCNC*, vol. 3, pp. 1540-1545, Mar. 2004.

[12] S. Soro and W. Heinzelman, "On the coverage problem in video-based wireless sensor networks," in *Proc. of International Conference on Broadband Networks*, vol. 2, pp. 932-939, Oct. 2005.

[13] E. Gilbert, "Capacity of a burst-noise channel," *Bell Systems Technical Journal*, vol. 39, no. 9, pp. 1253-1265, Sep. 1960.

[14] S. Mao, Y. T. Hou, X. Cheng, H. D. Sherali, and S. F. Midkiff, "Multipath routing for multiple description video over wireless ad hoc networks," in *Proc. of IEEE INFOCOM*, pp. 740-750, Mar. 2005.

[15] H. Wu and A. Abouzeid, "Error robust image transmission in wireless sensor networks," in *Proc. of the 5th Workshop on ASWN*, Jun. 2005.

[16] Z. He, Y. Liang, L. Chen, I. Ahmad, and D.Wu, "Power-rate-distortion analysis for wireless video communication under energy constraint," *IEEE Transactions on Circuits and Systems for Video Technology*, vol. 15, no. 5, pp. 645-658, May 2005.

[17] W. Heinzelman, A. Chandrakasan, and H. Balakrishnan, "An application-specific protocol architecture for wireless microsensor networks," *IEEE Transactions on Wireless Communications*, vol. 1, no. 4, pp. 660-670, Oct. 2002.

[18] T. S. Rappaport, *Wireless Communications: Principles and Practice, 2nd* edition, Prentice Hall, 2002.

[19] S. Boyd and L. Vandenberghe, *Convex Optimization*, Cambridge University Press, 2004.

[20] D. Palomar and M. Chiang, "A tutorial on decomposition methods and distributed network resource allocation," *IEEE Journal on Selected Areas in Communications*, vol. 24, no. 8, pp. 1439-1451, Aug. 2006.

[21] D. P. Bertsekas, A. Nedic, and A. E. Ozdaglar, *Convex Analysis and Optimization*, Athena Scientific, 2003.

[22] M. Busse, T. Haenselmann, T. King, and W. Effelsberg, "The impact of forward error correction on wireless sensor network performance," in *Proc. of ACM RealWSN*, May 2006.

[23] I. Reed and G. Solomon, "Polynomial codes over certain finite fields," *SIAM Journal on Applied Mathematics*, vol. 8, no. 2, pp. 300-304, 1960.

[24] H. Schulzrinne, S. Casner, R. Frederick, and V. Jacobson, "RTP: a transport protocol for real-time applications," *RFC 3550, Internet Engineering Task Force*, Jul. 2003.

[25] S. Appadwedula, M. Goel, N. R. Shanbhag, D. L. Jones, and K. Ramchandran, "Total system energy minimization for wireless image transmission," *Journal of VLSI Signal Processing Systems,* vol. 27, no. 1, pp. 99-117, Feb. 2001.

[26] Y. He, I. Lee, and L. Guan, "Optimized multi-path routing using dual decomposition for wireless video streaming," in *Proc. of IEEE ISCAS,* pp. 977-980, May 2007.

[27] Z. He, J. Cai, and C. W. Chen, "Joint source channel rate-distortion analysis for adaptive mode selection and rate control in wireless video coding," *IEEE Transactions on Circuits and Systems for Video Technology,* vol. 12, no. 6, pp. 511-523, Jun. 2002.

[28] Y. He, I. Lee, and L. Guan, "Distributed algorithms for network lifetime maximization in wireless visual sensor networks," *IEEE Transactions on Circuits and Systems for Video Technology,* vol. 19, no. 5, pp. 704-718, May 2009.

Index